Algebraic Theory
of Automata and Languages

Algebraic Theory of Automata and Languages

Masami Ito
Kyoto Sangyo University, Japan

World Scientific
NEW JERSEY • LONDON • SINGAPORE • BEIJING • SHANGHAI • HONG KONG • TAIPEI • CHENNAI

Published by

World Scientific Publishing Co. Pte. Ltd.
5 Toh Tuck Link, Singapore 596224
USA office: 27 Warren Street, Suite 401-402, Hackensack, NJ 07601
UK office: 57 Shelton Street, Covent Garden, London WC2H 9HE

British Library Cataloguing-in-Publication Data
A catalogue record for this book is available from the British Library.

ALGEBRAIC THEORY OF AUTOMATA AND LANGUAGES

ISBN-13 978-981-02-4727-0
ISBN-10 981-02-4727-3

In Memory of Professor Jürgen Duske

Preface

The theory of formal languages began with the classification of languages by N. Chomsky in Syntactic Structures in 1957. Now, this classification is called the Chomsky hierarchy of languages.

On the other hand, the theory of automata was initiated by M.O. Rabin and D. Scott in 1959. Their work can be regarded as the most important first step in the theory of automata in spite of its simplicity.

Since then, these two fields have been developed by many researchers as two important theoretical foundations of computer science.

In this book, we will mainly handle formal languages and automata from the algebraic point of view. In the first two chapters, we will investigate the algebraic structure of automata and then we will deal with a kind of global theory, i.e. partially ordered sets of automata. In the following four chapters, we will study grammars, languages and operations on languages. In the last section, we will introduce special kinds of automata, i.e. directable automata. The subjects in the book seem to be unique compared to other books with similar titles. The contents of the book are based on the author's work which started in the mid 1970s.

The author recognizes the importance of much research completed prior to the beginning of his own work. He would like to thank his co-authors of the joint papers which have become the basis of this book, i.e. Prof. G. Thierrin, Dr. S.S. Yu, Dr. L. Kari, Dr. P.V. Silva, Dr. B. Imreh, Dr. M. Katsura and Dr. M. Steinby. He would like to dedicate this book to Prof. J. Duske who passed away in 2000. In 1982, Prof. Duske invited the author to stay in Hannover for four months and offered the author the opportunity to do joint work with him. During this collaboration, the author was able to learn much from him from the points of view of mathematics and humanity. We were able to publish in a very short time two joint papers of which the author is very proud. One of these papers was a paper on directable automata which has become a foundation of this book as well.

The author is also grateful to Mr. T. Kadota for his assistance in making files of the manuscript of this book. The author appreciates Mr. Y. Kunimochi, Mr. H. Onoda and Dr. M. Toyama very much for their assistance in editing this book. He thanks also Mr. C. Everett for proofreading the English in the book.

The author thanks to Dr. Cs. Imreh and Dr. K. Tsuji for their careful reading of the manuscript and useful comments. Dr. Tsuji is the most recent of the author's co-author of a paper related to this book.

Finally, the author is thankful to the staff at the World Scientific Publishing Company for their assistance. Especially, he would like to express his gratitude to Dr. J.T. Lu for the extended patience and encouragement to the author.

January 2004 Masami Ito

Contents

Chapter 0

Introduction

Let A be a nonempty set. Throughout this book, by $|A|$ we denote the cardinality of A. Let A, B be sets and let ρ be a mapping of A into B. Then ρ is called a *surjection* if $\rho(A) = \{\rho(a) \mid a \in A\} = B$ and ρ is called an *injection* if, for any $a, a' \in A$, $a \neq a'$ if and only if $\rho(a) \neq \rho(a')$. If a mapping is a surjection and at the same time an injection, then it is called a *bijection*. An algebraic system $(A, \cdot)$ is called a *semigroup* if it satisfies the following conditions: (1) $a \cdot b \in A$ for any $a, b \in A$. (2) $a \cdot (b \cdot c) = (a \cdot b) \cdot c$ for any $a, b, c \in A$.

A semigroup A having the identity e is called a *monoid* where $a \cdot e = e \cdot a = a$ for any $a \in A$. Let $a, b \in A$. If there is no danger of confusion, we denote A and ab instead of $(A, \cdot)$ and $a \cdot b$.

A *group* G is a monoid having the inverse element g^{-1} of any $g \in G$ satisfying the condition: $gg^{-1} = g^{-1}g = e$. A subset $H \subseteq G$ is called a *subgroup* of a group G if H itself is a group. An important subgroup of a group is a normal subgroup. Let H be a subgroup of a group G. Then H is called a *normal subgroup* of G if $g^{-1}Hg \subseteq H$ holds for any $g \in G$. A normal subgroup induces a quotient group. Let $H \subseteq G$ be a normal subgroup of G. Then G can be decomposed into $G = H + g_1H + g_2H + \cdots$ where $g_i \in G, i \geq 1$ and $+$ stands for a disjoint union. For the set $\{H, g_1H, g_2H, \ldots\}$, we can introduce the multiplication operation as follows: (1) $H \cdot H = H$. (2) $H \cdot g_iH = g_iH \cdot H = g_iH$ for any $i \geq 1$. (3) $g_iH \cdot g_jH = g_ig_jH$ for any $i, j \geq 1$. Together with this operation, the set $\{H, g_1H, g_2H, \ldots\}$ forms a group, called the *quotient group*, and denoted by G/H. It is obvious that any group G has $\{e\}$ and G, itself as normal subgroups where e is the identity of G. A group G whose normal subgroups are only $\{e\}$ and G is called a *simple group*.

A semigroup (monoid, group) A is said to be *commutative* if $ab = ba$ for any $a, b \in A$. Let A and A' be monoids (groups) and let ρ be a mapping of A into A'. Then ρ is called a *homomorphism* if $\rho(ab) = \rho(a)\rho(b)$ and $\rho(e) = e'$ for any $a, b \in A$ where e and e' are the identities of A and A', respectively. Moreover, if ρ is a bijection, then ρ is called an *isomorphism*. In this case, $\rho(A) = A'$ and we denote $A \approx A'$. A homomorphism of a monoid into itself is called an *endomorphism* and an isomorphism of a monoid (group) onto itself is called *automorphism*.

Let ρ be a homomorphism of a group G onto a group G'. Then the set $\{g \in G \mid \rho(g) = e'$ where e' is the identity of $G'\}$ is called the *kernel* of ρ and denoted by $Ker(\rho)$. $Ker(\rho)$ is a normal subgroup of G and $G' \approx G/Ker(\rho)$. Conversely, if H is a normal subgroup of a group G, then we have the following homomorphism φ called the *natural homomorphism* of G onto G/H: $\varphi(g) = gH$ for any $g \in G$. Moreover, $H = Ker(\varphi)$.

Let X be a finite nonempty set. In formal language theory, X is called an *alphabet*. By X^+ we denote the set of all finite sequences of the elements of X. X^+ is called the *free semigroup* generated by X and it has the following operation, called the *concatenation*: For any $x, y \in X^+$, $xy \in X^+$.

Adjoining the identity ϵ to X^+, we obtain X^* which is called the *free monoid* generated by X. Any element of X^* is called a *word* over X and ϵ is called the *empty word*. Any subset of X^* is called a *language* over X. The study of words and languages is one of our main goals. There is a system which produces a language, called a *grammar*. A grammar is described by a quadruple $\mathcal{G} = (V, X, P, S)$. In this book, we will mainly deal with *regular grammars* and *context-free grammars*. Languages generated by regular grammars and context-free grammars are called *regular languages* and *context-free languages*, respectively.

Let $C \subseteq X^+$ and let $u_1, u_2, \ldots, u_r, v_1, v_2, \ldots, v_s \in C$ where r and s are positive integers. If $u_1u_2 \cdots u_r = v_1v_2 \cdots v_s$ implies that $r = s$ and $u_i = v_i$ for any $i = 1, 2, \ldots, r$, then C is called a *code* over X.

In this book, to investigate several properties of languages, we will deal with some codes, e.g. *prefix* codes, *suffix* codes, *bifix* codes, *infix* codes and *hypercodes*.

Another important notion is an automaton. A triple $\boldsymbol{A} = (S, X, \delta)$ is called an *automaton*. Moreover, a quintuple $\boldsymbol{A} = (S, X, \delta, s_0, F)$ whose initial triple (S, X, δ) constitutes an automaton is called an *acceptor* or a *finite acceptor*. Acceptors accept (or recognize) regular languages. We will also define a *pushdown acceptor* which accepts (or recognizes) a context-free language.

A relation $\leq$ on a set A is called a *partial order* if it satisfies the conditions: (1) $a \leq a$ for any $a \in A$. (2) $a = b$ if $a \leq b$ and $b \leq a$ for any $a, b \in A$. (3) $a \leq c$ if $a \leq b$ and $b \leq c$ for any $a, b, c \in A$. A set is called a *partially ordered set* (or in short, a *poset*) if it is followed by a partial order.

A poset is called a *lattice* if there exist a least upper bound and a greatest lower bound for any two elements.

Consider the following problem. Let $\mathcal{O}$ be a collection of objects, let $\mathcal{R}$ be a finite set of rules and let $\boldsymbol{P}$ be a property. If there is a procedure to decide whether or not any given $\boldsymbol{A}$ from $\mathcal{O}$ has property $\boldsymbol{P}$ after a finite time's usage of rules from $\mathcal{R}$, then the procedure is called an *algorithm* and the problem is called *decidable*. If there is no algorithm to decide the above problem, then the problem is called *undecidable*.

Let $\mathcal{O}$ be again a collection of objects and let $\mathcal{R}$ be a finite set of rules. Let $\boldsymbol{A}$ be any object from $\mathcal{O}$. If there is a procedure to construct $\boldsymbol{A}$ with a finite time's usage of rules from $\mathcal{R}$, then this procedure is also called an *algorithm* and the construction is said to be *effectively constructed*.

In this book, sometimes the logical symbols $\forall$ and $\exists$ will be used. For instance, "$\forall a \in A, \exists b \in B, a = f(b)$" means that "for any $a \in A$, there exists $b \in B$ such that $a = f(b)$ holds". On the other hand, "$\exists a \in A, \forall b \in B, a = f(b)$" means that "there exists $a \in A$ such that $a = f(b)$ holds for any $b \in B$".

As basic references, the following books are recommended: [12] and [13] for semigroups, [22] for groups, [15], [24], [23], [62] and [70] for automata, formal languages and acceptors in general, [19] for context-free languages, [8] and [65] for codes and [56] for combinatorial properties of words, and [9] for lattices.

This book consists of the following 9 chapters.

In Chapter 1, we will mainly deal with the automorphism groups of strongly connected automata and (n, G)-automata, i.e. representations of strongly connected automata.

In Chapter 2, we will genralize the results in Chapter 1 to the class of general automata.

In Chapter 3, we will consider partially ordered sets of automata where partial orders are induced by homomorphisms of automata.

In Chapter 4, we will deal with the compositions and decompositions of regular languages under *n-insertion* and *shuffle* operations. Moreover, we will consider a decidability problem with respect to the shuffle closures of regular commutative languages.

In Chapter 5, we will determine the structure of a shuffle closed language.

In Chapter 6, insertion and deletion operations will be treated in details.

In Chapter 7, shuffle and scattered deletion operations will be dealt with.

In Chapter 8, first we will provide the concept of *directable* automata and later we will deal with *nondeterministic directable* automata.

Chapter 1

Group-Matrix Type Automata

The study of endomorphism monoids and automorphism groups of automata was started by [17] and [68] and followed by [18], [69], [3], [67], [5], [6], [4], [63] and [57].

In this chapter, we will deal with a method to determine the structures of strongly connected automata whose automorphism groups are isomorphic to a given finite group.

For this purpose, we provide representations of strongly connected automata, called *regular group-matrix type* automata. In the first four sections, we will deal with correlations between strongly connected automata and regular group-matrix type automata. The last three sections contain applications of related results to the theory of automorphism groups of automata.

As for input sets of automata, we will avoid an abstract treatment. That is, we will restrict them to finite alphabets instead of general semigroups, so that the actions of automata will be described in the concrete form by means of state transition diagrams. However, our theory will be easily transferred to the general case.

The contents of this chapter are mainly based on the results in [31], [32], [34] and [35].

1.1 Strongly connected automata

In the present section, we provide some concepts on strongly connected automata and their automorphism groups, and present their fundamental results. Most of results were obtained by Fleck [17].

Definition 1.1.1 An *automaton* $\boldsymbol{A} = (S, X, \delta)$ consists of the following data: (1) S is a finite nonempty set, called a *state set.* (2) X is a finite nonempty set, called an *alphabet.* (3) δ is a function, called a *state transition function* of $S \times X$ into S.

The state transition function δ can be extended to the function of $S \times X^*$ into S as follows: (1) $\delta(s, \epsilon) = s$ for any $s \in S$. (2) $\delta(s, au) = \delta(\delta(s, a), u)$ for any $s \in S, a \in X$ and $u \in X^*$.

Definition 1.1.2 Let $\boldsymbol{A} = (S, X, \delta)$ and $\boldsymbol{B} = (T, X, \gamma)$ be automata and let ρ be a mapping of S into T. If $\rho(\delta(s, a)) = \gamma(\rho(s), a)$ holds for any $s \in S$ and any $a \in X$, then ρ is called a *homomorphism* of $\boldsymbol{A}$ into $\boldsymbol{B}$. If ρ is a bijection, then ρ is called an *isomorphism* of $\boldsymbol{A}$ onto $\boldsymbol{B}$. If there exists an isomorphism of $\boldsymbol{A}$ onto $\boldsymbol{B}$, then $\boldsymbol{A}$ and $\boldsymbol{B}$ are said to be *isomorphic to each other* and denoted by $\boldsymbol{A} \approx \boldsymbol{B}$. Moreover, if $\boldsymbol{A} = \boldsymbol{B}$, then a homomorphism is called an *endomorphism* and an isomorphism is called an *automorphism.* The set $E(\boldsymbol{A})$ ($G(\boldsymbol{A})$) of all endomorphisms (automorphisms) forms a monoid (group) and it is called the *endomorphism monoid* (*automorphism group*) of $\boldsymbol{A}$.

Now we define a special kind of automaton.

Definition 1.1.3 An automaton $\boldsymbol{A} = (S, X, \delta)$ is said to be *strongly connected* if for any pair of states $s, t \in S$ there exists an element $x \in X^*$ such that $\delta(s, x) = t$.

It can be easily seen that an endomorphism of a strongly connected automaton is surjective and hence bijective. Thus $E(\boldsymbol{A}) = G(\boldsymbol{A})$ for a strongly connected automaton $\boldsymbol{A}$.

Proposition 1.1.1 *If $\boldsymbol{A} = (S, X, \delta)$ is a strongly connected automaton and g, h are elements in $G(\boldsymbol{A})$ such that $g(s_0) = h(s_0)$ for some $s_0 \in S$, then $g(s) = h(s)$ for any $s \in S$.*

Proof Assume that $g(s_0) = h(s_0)$. Since $\boldsymbol{A}$ is stronly connected, there exists $x \in X^*$ such that $s = \delta(s_0, x)$. Hence $g(s) = g(\delta(s_0, x)) = \delta(g(s_0), x) = \delta(h(s_0), x) = h(\delta(s_0, x)) = h(s)$.

Proposition 1.1.2 *If $\boldsymbol{A} = (S, X, \delta)$ is a strongly connected automaton, then $|G(\boldsymbol{A})|$ divides $|S|$.*

Proof Let $s, t \in S$. If there exists some $g \in G(\boldsymbol{A})$ such that $t = g(s)$, then we write $s \sim t$. Then it can be shown that the relation $\sim$ is a congruence relation on S. Let $[s]$ be a congruence class containing $s \in S$, i.e. $[s] = \{g(s) \mid g \in G(\boldsymbol{A})\}$. By Proposition 1.1.1, $g(s) \neq h(s)$ for any $s \in S$ if $g, h \in G(\boldsymbol{A})$ and $g \neq h$. This means that $|[s]| = |G(\boldsymbol{A})|$. Notice that S is a disjoint union of $\{[s] \mid s \in S\}$. Hence $|G(\boldsymbol{A})|$ divides $|S|$

Definition 1.1.4 An automaton $\boldsymbol{A} = (S, X, \delta)$ is said to be *commutative* if $\delta(s, xy) = \delta(s, yx)$ for any $s \in S$ and any $x, y \in X^*$. If, in addition, $\boldsymbol{A}$ is strongly connected, then $\boldsymbol{A}$ is said to be *perfect.*

Proposition 1.1.3 *If $\boldsymbol{A} = (S, X, \delta)$ is a perfect automaton, then $G(\boldsymbol{A})$ is commutative and $|G(\boldsymbol{A})| = |S|$.*

Proof Let $x \in X^*$. By g_x, we denote the following mapping: $g_x(s) = \delta(s, x)$ for any $s \in S$. Let $s \in S$ and let $a \in X$. From the fact that $\boldsymbol{A}$ is commutative, it follows that $g_x(\delta(s, a)) = \delta(\delta(s, a), x) = \delta(s, ax) = \delta(s, xa) = \delta(\delta(s, x), a) = \delta(g_x(s), a)$. Hence $g_x \in E(\boldsymbol{A})$. Since $\boldsymbol{A}$ is strongly connected, $g_x \in G(\boldsymbol{A})$. Let $s, t \in S$. As $\boldsymbol{A}$ is strongly connected, there exists $x \in X^*$ such that $t = \delta(s, x)$, i.e. $t = g_x(s)$. This means that $s \sim t$ for any $s, t \in S$. Consequently, $G(\boldsymbol{A}) = \{g_x \mid x \in X^*\}$ and $|G(\boldsymbol{A})| = |S|$. Now let $g, h \in G(\boldsymbol{A})$. Then there exist $x, y \in X^*$ such that $g_x = g$ and $g_y = h$. Let $s \in S$. Then $g(s) = \delta(s, x)$ and $h(s) = \delta(s, y)$. Notice that $\boldsymbol{A}$ is commutative. Hence $gh(s) = \delta(h(s), x) = \delta(\delta(s, y), x) = \delta(s, xy) = \delta(s, yx) = \delta(\delta(s, y), x) = \delta(h(s), x) = gh(s)$, i.e. $gh = hg$. Thus $G(\boldsymbol{A})$ is commutative.

Proposition 1.1.4 *Let $\boldsymbol{A} = (S, X, \delta)$ be a strongly connected automaton. Then if $G(\boldsymbol{A})$ is commutative and $|G(\boldsymbol{A})| = |S|$, $\boldsymbol{A}$ is perfect.*

Proof Let $x, y \in X^*$ and let $s \in S$. Since $\boldsymbol{A}$ is strongly connected, and $|G(\boldsymbol{A})| = |S|$, there exist $g, h \in G(\boldsymbol{A})$ such that $g(s) = \delta(s, x)$ and $h(s) = \delta(s, y)$. Hence $\delta(s, xy) = \delta(\delta(s, x), y) = \delta(g(s), y) = g(\delta(s, y)) = gh(s) = hg(s) = h(\delta(s, x)) = \delta(h(s), x) = \delta(\delta(s, y), x) = \delta(s, yx)$, i.e. $\boldsymbol{A}$ is commutative. Therefore, $\boldsymbol{A}$ is perfect.

Definition 1.1.5 An automaton $\boldsymbol{A} = (S, X, \delta)$ is called a *permutation* automaton if $\delta(s, a)$ is a permutation on S for any $a \in X$.

Proposition 1.1.5 *Let $\boldsymbol{A} = (S, X, \delta)$ be a strongly connected automaton. Then if $|G(\boldsymbol{A})| = |S|$, $\boldsymbol{A}$ is a permutation automaton.*

Proof Let $s, t \in S$ with $s \neq t$. Since $|G(\boldsymbol{A})| = |S|$, there exists $g \in G(\boldsymbol{A})$ such that $t = g(s)$. Suppose $\delta(s, a) = \delta(t, a)$ for some $a \in X$. Then $\delta(s, a) = \delta(g(s), a) = g(\delta(s, a))$. By Proposition 1.1.1, g is the identity of $G(\boldsymbol{A})$, i.e. $t = s$, which is a contradiction. Therefore, $\delta(s, a) \neq \delta(t, a)$ for any $a \in X$. Thus $\boldsymbol{A}$ is a permutation automaton.

Proposition 1.1.6 *If $\boldsymbol{A}$ and $\boldsymbol{B}$ are automata such that $\boldsymbol{A} \approx \boldsymbol{B}$, then $G(\boldsymbol{A})$ is isomorphic to $G(\boldsymbol{B})$, i.e. $G(\boldsymbol{A}) \approx G(\boldsymbol{B})$.*

Proof Let $\boldsymbol{A} = (S, X, \delta)$ and $\boldsymbol{B} = (T, X, \gamma)$ be automata and let ρ be an isomorphism of $\boldsymbol{A}$ onto $\boldsymbol{B}$. First, we prove that ρ^{-1} is an isomorphism of $\boldsymbol{B}$ onto $\boldsymbol{A}$. Let $t \in T$ and let $a \in X$. Since ρ is surjective, there exists $s \in S$ such that $\rho(s) = t$, i.e. $s = \rho^{-1}(t)$. Notice that $\rho(\delta(s, a)) = \gamma(\rho(s), a)$. Hence $\rho(\delta(\rho^{-1}(t), a)) = \gamma(t, a)$. Thus $\delta(\rho^{-1}(t), a) = \rho^{-1}(\gamma(t, a))$. This means that ρ^{-1} is an isomorphism of $\boldsymbol{B}$ onto $\boldsymbol{A}$. Now we prove that $\rho g \rho^{-1} \in G(\boldsymbol{B})$ for any $g \in G(\boldsymbol{A})$. Let $t \in T$ and $a \in X$. Then $\rho g \rho^{-1}(\gamma(t, a)) = \rho g(\delta(\rho^{-1}(t), a)) = \rho(\delta(g\rho^{-1}(t), a)) = \gamma(\rho g \rho^{-1}(t), a)$. Therefore, we have $\rho g \rho^{-1} \in G(\boldsymbol{B})$. The correspondence $g \longleftrightarrow \rho g \rho^{-1}$ is one-to-one and hence $G(\boldsymbol{A}) \approx G(\boldsymbol{B})$.

Definition 1.1.6 An automaton $\boldsymbol{A} = (S, X, \delta)$ is said to be *simplified* if for any pair of inputs $a, b \in X$ with $a \neq b$ there exists an element $s \in S$ such that $\delta(s, a) \neq \delta(s, b)$.

1.2 Group-matrix type automata

In Section 1.3, we will give representations of strongly connected automata. This section provides the theoreical basis for Section 1.3.

Definition 1.2.1 Let G be a finite group. Then G^0 is the set $G \cup \{0\}$ in which we introduce two operations $\cdot$ and $+$:

(1) For any $g, h \in G$, we define $g \cdot h$ as the group operation in G.

(2) For any $g \in G$, we define $g \cdot 0 = 0 \cdot g = 0$ and $0 \cdot 0 = 0$.

(3) For any $g \in G$, we define $g + 0 = 0 + g = g$ and $0 + 0 = 0$.

(4) For any $g, h \in G$, $g + h$ is not defined.

We will use sometimes the notations gh and $\sum_{i=1}^{s} g_i$ instead of $g \cdot h$ and $g_1 + g_2 + \cdots + g_s$. Notice that the sum $\sum_{i=1}^{s} g_i$ is defined only if at most one of $g_i, i = 1, 2, \ldots, s$, is not zero.

Definition 1.2.2 Let G be a finite group and let n be a positive integer. We consider an $n \times n$ matrix (f_{pq}), $f_{pq} \in G^0, p, q = 1, 2, \ldots, n$. If an $n \times n$ matrix (f_{pq}) satisfies the following conditions, then (f_{pq}) is called a *group-matrix of order n on G*:

For any $p' = 1, 2, \ldots, n$, there exists a unique number $q' = 1, 2, \ldots, n$ such that $f_{p'q'} \neq 0$.

We denote by $\widetilde{G_n}$ the set of all group-matrices of order n on G. Then $\widetilde{G_n}$ forms a semigroup under the following operation:

$$(f_{pq})(g_{pq}) = \left(\sum_{k=1}^{n} f_{pk} g_{kq} \right).$$

Definition 1.2.3 Let G be a finite group and n be a positive integer. We consider a vector $(f_p), f_p \in G^0, p = 1, 2, \ldots, n$. A vector (f_p) is called a *group-vector of order n* on G, if there exists a unique number $p' = 1, 2, \ldots, n$ such that $f_{p'} \neq 0$. We denote by $\widehat{G_n}$ the set of all group-vectors of order n on G. For any $(f_p) \in \widehat{G_n}$ and any $(g_{pq}) \in \widetilde{G_n}$, we define the following multiplication:

$$(f_p)(g_{pq}) = \left(\sum_{k=1}^{n} f_k g_{kp}\right).$$

Under this operation, we have $(f_p)(g_{pq}) \in \widehat{G_n}$.

Definition 1.2.4 Let G be a finite group and n be a positive integer. An automaton $\boldsymbol{A} = (\widehat{G_n}, X, \delta_\Psi)$ is called a *group-matrix type automaton of order* n on G, or simply an (n, G)*-automaton*, if the following conditions are satisfied

(1) $\widehat{G_n}$ is the set of states.

(2) X is a set of inputs.

(3) δ_Ψ is a state transition function and it is defined by

$\delta_\Psi(\hat{g}, a) = \hat{g}\Psi(a), \hat{g} \in \widehat{G_n}$, $a \in X$ where Ψ is a mapping of X into $\widetilde{G_n}$.

Remark 1.2.1 The mapping Ψ can be extended to the mapping of X^* into $\widetilde{G_n}$ as follows:

> $\Psi(\epsilon) = (e_{pq})$, $e_{pq} = 0$ if $p \neq q$, and $e_{pp} = e$ where e is the identity of G, and $\Psi(au) = \Psi(a)\Psi(u)$ for any $a \in X$ and $u \in X^*$.

In this case, we can easily see that $\delta_\Psi(\hat{g}, x) = \hat{g}\Psi(x)$ holds for any $x \in X^*$.

The following two results are obvious:

Proposition 1.2.1 *Let* $\boldsymbol{A} = (\widehat{G_n}, X, \delta_\Psi)$ *be an* (n, G)*-automaton. Then* $\boldsymbol{A}$ *is simplified if and only if* Ψ *is a injective mapping of* X *into* $\widetilde{G_n}$.

Proposition 1.2.2 *Let* $\boldsymbol{A}$ *be a* $(1, G)$*-automaton. Then* $\boldsymbol{A}$ *is a permutation automaton.*

Example 1.2.1 $G = \{e, g\}, e = g^2$, $X = \{a, b\}$, $\Psi(a) = \begin{pmatrix} g & 0 \\ 0 & e \end{pmatrix}$, $\Psi(b) = \begin{pmatrix} 0 & g \\ g & 0 \end{pmatrix}$, $\boldsymbol{A} = (\widehat{G_2}, X, \delta_\Psi)$.

$\widehat{G_2} = \{s_1, s_2, s_3, s_4\}$, $s_1 = (e, 0)$, $s_2 = (g, 0)$, $s_3 = (0, e)$, $s_4 = (0, g)$.

$\delta_\Psi(s_1, a) = s_1\Psi(a) = (e, 0)\begin{pmatrix} g & 0 \\ 0 & e \end{pmatrix} = (g, 0) = s_2$, $\delta_\Psi(s_1, b) = s_4$, $\delta_\Psi(s_2, a) = s_1$, $\delta_\Psi(s_2, b) = s_3$, $\delta_\Psi(s_3, a) = s_3$, $\delta_\Psi(s_3, b) = s_2$, $\delta_\Psi(s_4, a) = s_4$, $\delta_\Psi(s_4, b) = s_1$.

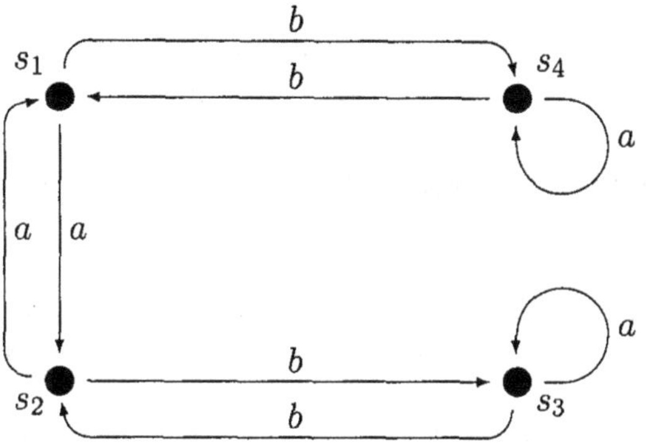

Figure 1.1: State transition diagram of $\boldsymbol{A}$

Theorem 1.2.1 *Let $\boldsymbol{A} = (\widehat{G_n}, X, \delta_\Psi)$ be an (n, G)-automaton. Then G is isomorphic to a subgroup of $G(\boldsymbol{A})$.*

Proof For any $g \in G$, we will define the following mapping ρ_g of $\widehat{G_n}$ onto itself: $\rho_g(\hat{h}) = (gh_p), p = 1, 2, \ldots, n$ for any $\hat{h} \in \widehat{G_n}$, where $\hat{h} = (h_p), h_p \in G^0, p = 1, 2, \ldots, n$.

By the definition, we can easily see that ρ_g is a permutation on $\widehat{G_n}$.

Now, we prove that $\rho_g(\delta_\Psi(\hat{h}, a)) = \delta_\Psi(\rho(\hat{h}), a)$ for any $\hat{h} \in \widehat{G_n}$ and any $a \in X$. To this end, put $\hat{h} = (h_p), h_p \in G^0, p = 1, 2, \ldots, n$. Then we have: $\rho_g(\delta_\Psi(\hat{h}, a)) = \rho_g(\hat{h}\Psi(a)) = \rho_g((h_p)\Psi(a)) = (gh_p)\Psi(a) = \rho_g(\hat{h})\Psi(a) = \delta_\Psi(\rho_g(\hat{h}), a)$.

Therefore, we have $\rho_g \in G(\boldsymbol{A})$.

Next, we prove that the mapping $g \longrightarrow \rho_g$ is a homomorphism. For this purpose, put $g \rightarrow \rho_g$, $g' \rightarrow \rho_{g'}$, and $\hat{h} = (h_p)$. Then we have:

$$\rho_{gg'}(\hat{h}) = ((gg')h_p) = (gg'h_p) = \rho_g((g'h_p)) = \rho_g(\rho_{g'}(\hat{h})) = \rho_g\rho_{g'}(\hat{h}).$$

This means that $gg' \longrightarrow \rho_g\rho_{g'}$ holds, i.e. the mapping $g \longrightarrow \rho_g$ is a homomorphism.

We can easily see that the mapping $g \longrightarrow \rho_g$ is a surjective mapping. Therefore, the mapping $g \longrightarrow \rho_g$ is an isomorphism of G onto a subgroup of $G(\boldsymbol{A})$.

Definition 1.2.5 An (n, G)-automaton $\boldsymbol{A}$ is said to be *regular* if $\boldsymbol{A}$ is strongly connected and $G(\boldsymbol{A}) \approx G$ holds.

Concerning the strong connectedness, we have the following result:

Theorem 1.2.2 *An (n, G)-automaton $\boldsymbol{A} = (\widehat{G_n}, X, \delta_\Psi)$ is strongly connected if and only if the following condition is satisfied:*

> *For any $p', q' = 1, 2, \ldots, n$ and any $g \in G$, there exists some element x in X^* such that $\psi_{p'q'}(x) = g$ where $\Psi(x) = (\psi_{pq}(x))$.*

Proof First, assume that $\boldsymbol{A}$ is strongly connected. Put $\widehat{e_{p'}} = (e_{pp'}) = (0, \ldots, 0, \overset{p'\text{-th}}{e}, 0, \ldots, 0) \in \widehat{G_n}$ where e is the identity of G (for the notation, see Remark 1.2.1).

Now, we put $\widehat{g_{q'}} = (ge_{pq'}) = (0, \ldots, 0, \overset{q'\text{-th}}{g}, 0, \ldots, 0) \in \widehat{G_n}$ for any $g \in G$. Since $\boldsymbol{A}$ is strongly connected, there exists some element $x \in X^*$ such that $\delta_\Psi(\widehat{e_{p'}}, x) = \widehat{g_{q'}}$. Thus we have $\psi_{p'q'}(x) = g$.

Conversely, assume that the condition is satisfied. For any pair of elements $\hat{g}, \hat{h} \in \widehat{G_n}$, there exist some elements $g', h' \in G$ and some numbers $p', q' = 1, 2, \ldots, n$ such that $\hat{g} = (g'e_{pp'})$ and $\hat{h} = (h'e_{pq'})$. By the assumption, there exists some element $x \in X^*$ such that $\psi_{p'q'}(x) = g'^{-1}h'$. In this case, $\delta_\Psi(\hat{g}, x) = \hat{g}\Psi(x) = \hat{h}$ holds. Therefore, $\boldsymbol{A}$ is strongly connected.

First, we will give a necessary and sufficient condition on $\Psi(X)$ or $\Psi(X^*)$ in order that a group-matrix type automaton $\boldsymbol{A} = (\widehat{G_n}, X, \delta_\Psi)$ may be regular in the cases $n = 1$ and $n = 2$. The general case and the case that n is a prime number will be also treated.

Theorem 1.2.3 *Let $\boldsymbol{A} = (\widehat{G_1}, X, \delta_\Psi)$ be a $(1, G)$-automaton. Then if $\boldsymbol{A}$ is strongly connected, $\boldsymbol{A}$ is regular.*

Proof We have to prove that $G(\boldsymbol{A})$ is isomorphic to G. To this end, it is enough to show that for any $\rho \in G(\boldsymbol{A})$ there exists some element $g \in G$ such that $\rho = \rho_g$ (for the reason and the notation, see Theorem 1.2.1).

Assume $\rho \in G(\boldsymbol{A})$. Then we have $\rho(\delta_\Psi(\hat{h}, x)) = \delta_\Psi(\rho(\hat{h}), x)$ for any $\hat{h} = (h) \in \widehat{G_1}, h \in G$ and all $x \in X^*$. This means that $\rho(\hat{h}\Psi(x)) = \rho(\hat{h})\Psi(x)$ holds for any $\hat{h} = (h) \in G_1, h \in G$ and any $x \in X^*$.

Notice that for $\hat{e} = (e) \in \widehat{G_1}$ there exists some element $g \in G$ such that $\rho(\hat{e}) = (g) \in \widehat{G_1}$ where e is the identity of G. By the strong connectedness of $\boldsymbol{A}$, for any $h \in G$ there exists some element $x \in X^*$ such that $\Psi(x) = (h)$. Thus for any $\hat{h} = (h) \in \widehat{G_1}$, we have the following:

$$\rho(\hat{h}) = \rho(\hat{e}\Psi(x)) = \rho(\hat{e})\Psi(x) = (g)(h) = (gh) = \rho_g(\hat{h}).$$

Therefore, we have $\rho = \rho_g$.

Theorem 1.2.4 *Let $\boldsymbol{A} = (\widehat{G_2}, X, \delta_\Psi)$ be a strongly connected $(2, G)$-automaton. Then $\boldsymbol{A}$ is not regular if and only if there exist an automorphism φ of G, an element k in G, and two subsets $\Lambda, \Gamma, \Gamma \neq \emptyset$ of G such that $\varphi(k) = k$, $\varphi^2(g) = kgk^{-1}$ for any $g \in G$, and*

$$\Psi(X) = \left\{ \begin{pmatrix} g & 0 \\ 0 & \varphi(g) \end{pmatrix}, \begin{pmatrix} 0 & h \\ \varphi(h)k & 0 \end{pmatrix} \middle| g \in \Lambda, h \in \Gamma \right\}.$$

Proof Assume that $\boldsymbol{A}$ is not regular. Then there exists some element $\rho \in G(\boldsymbol{A})$ such that $\rho \neq \rho_g$ for any $g \in G$ (see Theorem 1.2.1). Moreover, there exist two transformations ρ_1, ρ_2 on G such that $\rho((g, 0)) = (0, \rho_2(g))$ and $\rho((0, h)) = (\rho_1(h), 0)$ for any $g, h \in G$. Because, if it is not true, there exist some elements $g', h' \in G$ such that either $\rho((g',0)) = (h',0)$ or $\rho((0,g')) = (0,h')$ holds. Consider the former case. In this case, since $\rho(\delta_\Psi((g',0),x)) = \delta_{\dot{\Psi}}(\rho((g',0)),x)$ holds, we have $\rho((g',0)\Psi(x)) = \rho((g',0))\Psi(x) = (h',0)\Psi(x)$ for any $x \in X^*$. By the strong connectedness of $\boldsymbol{A}$, we can find an element $x' \in X^*$ such that $\Psi(x') = \begin{pmatrix} g'^{-1}g & 0 \\ u & v \end{pmatrix}$, $u, v \in G^0$ and $u+v \in G$ for any $g \in G$. Substituting this value into the equality $\rho((g',0)\Psi(x')) = (h',0)\Psi(x')$, we have $\rho((g,0)) = (h'g'^{-1}g, 0)$ for any $g \in G$.

In the same manner, we can find an element $x'' \in X^*$ such that $\Psi(x'') = \begin{pmatrix} 0 & g'^{-1}h \\ u' & v' \end{pmatrix}$, $u', v' \in G^0$ and $u' + v' \in G$ for any $h \in G$. By the equality $\rho((g',0)\Psi(x'')) = (h',0)\Psi(x'')$, we have $\rho((0,h)) = (0, h'g'^{-1}h)$ for any $h \in G$. Thus ρ must be equal to $\rho_{h'g'^{-1}}$. This is a contradiction. Hence there exist two transformations ρ_1, ρ_2 on G.

We consider an element in $\Psi(X^*)$ of the form $\begin{pmatrix} g & 0 \\ u & v \end{pmatrix}$ where $u, v \in G^0$ and $u + v, g \in G$. Since $(e,0)\begin{pmatrix} g & 0 \\ u & v \end{pmatrix} = (g,0)$ and $\rho\left((e,0)\begin{pmatrix} g & 0 \\ u & v \end{pmatrix}\right) = \rho((e,0))\begin{pmatrix} g & 0 \\ u & v \end{pmatrix}$, we have $(0,\rho_2(g)) = (0,\rho_2(e))\begin{pmatrix} g & 0 \\ u & v \end{pmatrix} = (\rho_2(e)u, \rho_2(e)v)$. Therefore, $u = 0$ and $v = \rho_2(e)^{-1}\rho_2(g)$ holds. This means that, if $\begin{pmatrix} g & 0 \\ u & v \end{pmatrix} \in \Psi(X^*)$ holds, then we have $\begin{pmatrix} g & 0 \\ u & v \end{pmatrix} = \begin{pmatrix} g & 0 \\ 0 & \rho_2(e)^{-1}\rho_2(g) \end{pmatrix}$.

Now, consider an element in $\Psi(X^*)$ of the form $\begin{pmatrix} 0 & g \\ u & v \end{pmatrix}$ where $u, v \in G^0$ and $u + v, g \in G$. Since $(e,0)\begin{pmatrix} 0 & g \\ u & v \end{pmatrix} = (0,g)$ and $\rho\left((e,0)\begin{pmatrix} 0 & g \\ u & v \end{pmatrix}\right) = \rho((e,0))\begin{pmatrix} 0 & g \\ u & v \end{pmatrix}$, we have $(\rho_1(g),0) = (0,\rho_2(e))\begin{pmatrix} 0 & g \\ u & v \end{pmatrix} = (\rho_2(e)u, \rho_2(e)v)$. Therefore, $u = \rho_2(e)^{-1}\rho_1(g)$ and

$v = 0$ hold. This means that, if $\begin{pmatrix} 0 & g \\ u & v \end{pmatrix} \in \Psi(X^*)$ holds, then we have $\begin{pmatrix} 0 & g \\ u & v \end{pmatrix} = \begin{pmatrix} 0 & g \\ \rho_2(e)^{-1}\rho_1(g) & 0 \end{pmatrix}$.

From these facts and the strong connectedness of $\boldsymbol{A}$, we have the following:

> There exist two permutations φ, ξ on G such that for any $g, h \in G$ we have that $\begin{pmatrix} g & 0 \\ 0 & \varphi(g) \end{pmatrix}, \begin{pmatrix} 0 & h \\ \xi(h) & 0 \end{pmatrix} \in \Psi(X^*)$ and that conversely all elements in $\Psi(X^*)$ can be represented in the above form.

Since $\Psi(X^*)$ is a semigroup, we have the following four equalities.

By $\begin{pmatrix} g & 0 \\ 0 & \varphi(g) \end{pmatrix} \begin{pmatrix} h & 0 \\ 0 & \varphi(h) \end{pmatrix} = \begin{pmatrix} gh & 0 \\ 0 & \varphi(g)\varphi(h) \end{pmatrix}$,

(A) $\qquad \varphi(gh) = \varphi(g)\varphi(h)$ for any $g, h \in G$.

By $\begin{pmatrix} g & 0 \\ 0 & \varphi(g) \end{pmatrix} \begin{pmatrix} 0 & h \\ \xi(h) & 0 \end{pmatrix} = \begin{pmatrix} 0 & gh \\ \varphi(g)\xi(h) & 0 \end{pmatrix}$,

(B) $\qquad \xi(gh) = \varphi(g)\xi(h)$ for any $g, h \in G$.

By $\begin{pmatrix} 0 & g \\ \xi(g) & 0 \end{pmatrix} \begin{pmatrix} h & 0 \\ 0 & \varphi(h) \end{pmatrix} = \begin{pmatrix} 0 & g\varphi(h) \\ \xi(g)h & 0 \end{pmatrix}$,

(C) $\qquad \xi(g)h = \xi(g\varphi(h))$ for any $g, h \in G$.

By $\begin{pmatrix} 0 & g \\ \xi(g) & 0 \end{pmatrix} \begin{pmatrix} 0 & h \\ \xi(h) & 0 \end{pmatrix} = \begin{pmatrix} g\xi(h) & 0 \\ 0 & \xi(g)h \end{pmatrix}$,

(D) $\qquad \xi(g)h = \varphi(g\xi(h))$ for any $g, h \in G$.

Now we show that there exist k, Λ and Γ which satisfy the conditions in the theorem. Recall that we have already proven the existence of an automorphism φ of G.

First, $\Gamma \neq \emptyset$ because there exist the transformations ρ_1 and ρ_2. By (A), φ is an automorphism of G. Let $k = \xi(e)$. From (D), it follows that $\varphi(k) = \varphi(\xi(e)) = \varphi(e\xi(e)) = \xi(e)e = \xi(e) = k$. Let $g \in G$. Then by (B), we have $\xi(g) = \varphi(g)\xi(e) = \varphi(g)k$ and, by (D) we have $\varphi(\xi(g)) = \xi(e)g = kg$. Hence $kg = \varphi(\xi(g)) = \varphi(\varphi(g)k) = \varphi^2(g)\varphi(k) = \varphi^2(g)k$ and $\varphi^2(g) = kgk^{-1}$.

Conversely, assume that there exist φ, k, Λ and Γ which satisfy the conditions in the theorem. Put $\rho((g,0)) = (0,\varphi(g))$ and $\rho((0,h)) = (\varphi(h)k, 0)$ for any $g,h \in G$.

Let $g \in G$ and let $a \in X$. Assume that $\Psi(a) = \begin{pmatrix} h & 0 \\ 0 & \varphi(h) \end{pmatrix}$ for some $h \in G$. Then $\rho(\delta_\Psi((g,0),a)) = \rho((g,0)\begin{pmatrix} h & 0 \\ 0 & \varphi(h) \end{pmatrix}) = \rho((gh,0)) = (0,\varphi(gh)) = (0,\varphi(g)\varphi(h)) = (0,\varphi(g))\begin{pmatrix} h & 0 \\ 0 & \varphi(h) \end{pmatrix} = \delta_\Psi(\rho((g,0)),a)$.

Moreover, we have $\rho(\delta_\Psi((0,g),a)) = \rho((0,g)\begin{pmatrix} h & 0 \\ 0 & \varphi(h) \end{pmatrix}) = \rho((0,g\varphi(h))) = (\varphi(g\varphi(h))k,0) = (\varphi(g)\varphi^2(h)k,0) = (\varphi(g)khk^{-1}k,0)$ $= (\varphi(g)kh,0) = (\varphi(g)k,0)\begin{pmatrix} h & 0 \\ 0 & \varphi(h) \end{pmatrix} = \delta_\Psi(\rho((0,g)),a)$.

Assume that $\Psi(a) = \begin{pmatrix} 0 & h \\ \varphi(h)k & 0 \end{pmatrix}$ for some $h \in G$. Then we have $\rho(\delta_\Psi((g,0),a)) = \rho((g,0)\Psi(a)) = \rho((g,0)\begin{pmatrix} 0 & h \\ \varphi(h)k & 0 \end{pmatrix}) = \rho((0,gh)) = (\varphi(gh)k,0) = (\varphi(g)\varphi(h)k,0) = (0,\varphi(g))\begin{pmatrix} 0 & h \\ \varphi(h)k & 0 \end{pmatrix}$ $= \delta_\Psi(\rho((g,0)),a)$.

Furthermore, we have $\rho(\delta_\Psi((0,g),a)) = \rho((0,g)\Psi(a)) = \rho((0,g)$ $\begin{pmatrix} 0 & h \\ \varphi(h)k & 0 \end{pmatrix}) = \rho((g\varphi(h)k,0)) = (0,\varphi(g\varphi(h)k)) = (0,\varphi(g)\varphi^2(h)\varphi$ $(k)) = (0,\varphi(g)khk^{-1}k) = (0,\varphi(g)kh) = (\varphi(g)k,0)\begin{pmatrix} 0 & h \\ \varphi(h)k & 0 \end{pmatrix} = \delta_\Psi(\rho((0,g)),a)$.

Thus $\rho \in G(\boldsymbol{A})$. However, since $\rho \neq \rho_g$ for any $g \in G$, $\boldsymbol{A}$ is not regular.

Example 1.2.2 The automaton $\boldsymbol{A}$ mentioned in Example 1.2.1 is regular. The reason is obvious from Theorem 1.2.4.

Example 1.2.3 The following automaton $\boldsymbol{A}$ is strongly connected, but not regular.

$G = \{e, g\}$, $e = g^2$ and e is the identity of G, $X = \{a, b\}$, $\Psi(a) = \begin{pmatrix} g & 0 \\ 0 & g \end{pmatrix}$, $\Psi(b) = \begin{pmatrix} 0 & g \\ g & 0 \end{pmatrix}$, $\boldsymbol{A} = (\widehat{G_2}, X, \delta_\Psi)$. $\widehat{G_2} = \{s_1, s_2, s_3, s_4\}$, $s_1 = (e, 0)$, $s_2 = (g, 0)$, $s_3 = (0, e)$, $s_4 = (0, g)$.

$\delta_\Psi(s_1, a) = s_2$, $\delta_\Psi(s_1, b) = s_4$, $\delta_\Psi(s_2, a) = s_1$, $\delta_\Psi(s_2, b) = s_3$, $\delta_\Psi(s_3, a) = s_4$, $\delta_\Psi(s_3, b) = s_2$, $\delta_\Psi(s_4, a) = s_3$, $\delta_\Psi(s_4, b) = s_1$.

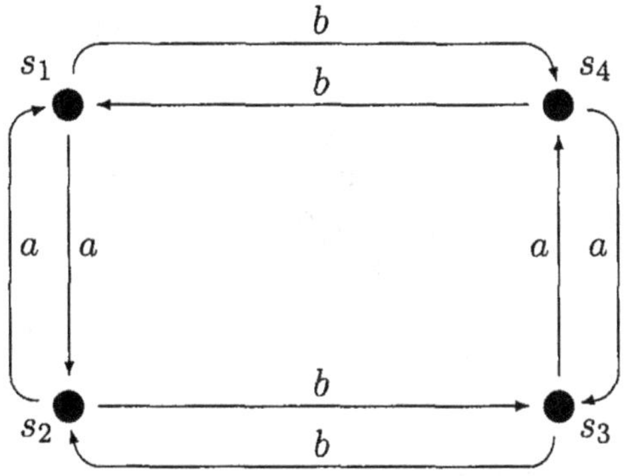

Figure 1.2: State transition diagram of $\boldsymbol{A}$

In the case $n > 2$, we have the following result. The proof of the necessity is due to [55].

Theorem 1.2.5 *Let $\boldsymbol{A} = (\widehat{G_n}, X, \delta_\Psi)$ be a strongly connected (n, G)-automaton. Then $\boldsymbol{A}$ is regular if and only if there exists a number $i' = 1, 2, \ldots, n$ which satisfies the following condition:*

> *For any $i = 1, 2, \ldots, n, i \neq i'$, there exist some elements $x, y \in X^*$ and number $q' = 1, 2, \ldots, n$ such that $\psi_{i'q}(x) = \psi_{i'q}(y)$ for any $q = 1, 2, \ldots, n$ and that $\psi_{iq'}(x) \neq \psi_{iq'}(y)$ where $\Psi(x) = (\psi_{pq}(x))$ and $\Psi(y) = (\psi_{pq}(y))$.*

Proof ($\Rightarrow$) Assume that the condition is not satisfied. Then there exist $i, j = 1, 2, \ldots, n, i \neq j$ such that for any $x, y \in X^*, \psi_{iq}(x) = \psi_{jq}(y)$ if and only if $\psi_{jq}(x) = \psi_{jq}(y)$ for any $q = 1, 2, \ldots, n$. We define a mapping ρ as follows: $\rho(\widehat{e_i}\Psi(x)) = \widehat{e_j}\Psi(x)$ for any $x \in X^*$.

Notice that, by the assumption, ρ is well defined as a mapping. Moreover, since $\boldsymbol{A}$ is strongly connected, ρ is a bijection of $\widehat{G_n}$ onto itself. Now we show that ρ is an automorphism of $\boldsymbol{A}$.

Let $\hat{g} \in \widehat{G_n}$. Since $\boldsymbol{A}$ is strongly connected, there exists $z \in X^*$ such that $\hat{g} = \widehat{e_i}\Psi(z)$. Let $a \in X$. Then $\rho(\delta_\Psi(\hat{g}, a)) = \rho(\hat{g}\Psi(a)) = \rho(\widehat{e_i}\Psi(z)\Psi(a)) = \rho(\widehat{e_i}\Psi(za)) = \widehat{e_j}\Psi(za) = \widehat{e_j}\Psi(z)\Psi(a) = \rho(\hat{g})\Psi(a) = \delta_\Psi(\rho(\hat{g}), a)$. Consequently, $\rho \in G(\boldsymbol{A})$. On the other hand, $\rho(\widehat{e_i}) = \widehat{e_j}$. Hence $\rho \neq \rho_g$ for any $g \in G$. Thus $\boldsymbol{A}$ is not regular. This means that the condition holds if $\boldsymbol{A}$ is regular.

($\Leftarrow$) It is enough to show that for any $\rho \in G(\boldsymbol{A})$ there exists some element $g \in G$ such that $\rho = \rho_g$. Therefore, we assume that $\rho \in G(\boldsymbol{A})$. Hence for any $\hat{h} \in \widehat{G_n}$ and any $x \in X^*$, $\rho(\hat{h})\Psi(x) = \rho(\hat{h}\Psi(x))$ holds. Now, put $\hat{h} = (e_{pi'})$. Then there exist some element $k \in G$ and number $i = 1, 2, \ldots, n$ such that $\rho(\hat{h}) = (ke_{pi})$.

Assume that $i = i'$. Let $j = 1, 2, \ldots, n$ and let $g \in G$. Since $\boldsymbol{A}$ is strongly connected, there exists $x \in X^*$ such that $\psi_{ij}(x) = g$ where $\Psi(x) = (\psi_{pq}(x))$. Then $\delta_\Psi(\hat{h}, x) = \hat{h}\Psi(x) = (ge_{pj})$. Therefore, $\rho(ge_{pj}) = \rho(\delta_\Psi(\hat{h}, x)) = \delta_\Psi(\rho(\hat{h}, x)) = \rho(\hat{h})\Psi(x) = (kge_{pj})$. Thus $\rho = \rho_k$.

Now, assume that $i \neq i'$. Then by a condition in the theorem, there exist some elements $x, y \in X^*$ and number $q' = 1, 2, \ldots, n$ such that $\psi_{i'q}(x) = \psi_{i'q}(y)$ for any $q = 1, 2, \ldots, n$ and that $\psi_{iq'}(x) \neq \psi_{iq'}(y)$. By the above relationship and by the equalities $\rho(\hat{h})\Psi(x) = \rho(\hat{h}\Psi(x))$, $\rho(\hat{h})\Psi(y) = \rho(\hat{h}\Psi(y))$, $\hat{h} = (e_{pi'})$ and $\rho(\hat{h}) = (ke_{pi}), i \neq i'$, the following contradiction follows:

$$\rho(\hat{h}\Psi(x)) \neq \rho(\hat{h}\Psi(y)), \text{ though } \hat{h}\Psi(x) = \hat{h}\Psi(y).$$

Therefore, $i = i'$ must hold and thus $\boldsymbol{A}$ is regular.

1.3 Representations of automata

In this section, we will give representations of strongly connected automata by group-matrix type automata. As preparation, the following lemma is provided:

Lemma 1.3.1 *Let G and G' be two isomorphic groups. Moreover, assume that $\boldsymbol{A} = (\widehat{G_n}, X, \delta_\Psi)$ is an (n, G)-automaton. Then there exists an (n, G')-automaton $\boldsymbol{A}' = (\widehat{G'_n}, X, \delta_{\Psi'})$ such that $\boldsymbol{A} \approx \boldsymbol{A}'$. If, in addition, $\boldsymbol{A}$ is regular, then $\boldsymbol{A}'$ is also regular.*

Proof Let Φ be an isomorphism of G onto G'. We define $\Psi'(a) = (\tilde{\Phi}(\psi_{pq}(a)))$ for any $a \in X$ where $\tilde{\Phi}$ is the extension of Φ on G^0 with $\tilde{\Phi}(0) = 0$ and $\Psi(a) = (\psi_{pq}(a))$.

With the above Ψ', we define $\boldsymbol{A}' = (\widehat{G'_n}, X, \delta_{\Psi'})$.

First, we prove that $\boldsymbol{A} \approx \boldsymbol{A}'$. To this end, we put $\rho(\hat{g}) = (\tilde{\Phi}(g_p))$ for any $\hat{g} = (g_p) \in \widehat{G_n}$. Then ρ is a bijective mapping of $\widehat{G_n}$ onto $\widehat{G'_n}$. By the definition of Ψ' and the fact that Φ is an isomorphism of G onto G', it is obvious that $\rho(\delta_\Psi(\hat{g}, a)) = \delta_{\Psi'}(\rho(\hat{g}), a)$. This means that $\boldsymbol{A} \approx \boldsymbol{A}'$.

Next, assume that $\boldsymbol{A}$ is regular. Then $G(\boldsymbol{A}) \approx G$ holds. By Proposition 1.1.6, we have $G(\boldsymbol{A}) \approx G(\boldsymbol{A}')$. On the other hand, $G \approx G'$ holds because of the assumption. Therefore, we have $G(\boldsymbol{A}') \approx G'$. This means that $\boldsymbol{A}'$ is regular.

Theorem 1.3.1 *Let $\boldsymbol{A} = (S, X, \delta)$ be a strongly connected automaton such that $|S| = n|G(\boldsymbol{A})|$ where n is a positive integer. Moreover, assume that G is a finite group such that $G \approx G(\boldsymbol{A})$. Then there exists a regular (n, G)-automaton which is isomorphic to $\boldsymbol{A}$.*

Proof By Lemma 1.3.1, it is enough to show that $\boldsymbol{A}$ is isomorphic to a regular $(n, G(\boldsymbol{A}))$-automaton. As preparation to prove this, we recall the following relation on S:

> For $s, t \in S$, $s \sim t$ means that there exists some element $g \in G(\boldsymbol{A})$ such that $t = g(s)$.

Since $\sim$ is an equivalence relation on S, it induces n equivalence classes.

Let $S_i, i = 1, 2, \ldots, n$ be an equivalence class by the relation $\sim$ on S. We choose $s_i \in S_i, i = 1, 2, \ldots, n$ and put $T = \{s_1, s_2, \ldots, s_n\}$. Then for any $a \in X$ and any $s_p \in T$, there exists a unique pair $s_{p'} \in T$, $h \in G(\boldsymbol{A})$ such that $\delta(s_p, a) = h(s_{p'})$. Making use of these p, p' and h, we define the following mapping Ψ of X into $\widetilde{G(\boldsymbol{A})}_n$:

$$\Psi(a) = (\psi_{pq}(a)) \,\text{for}\, a \in X \text{where}\; \psi_{pp'}(a) = h \,\text{and}\, \psi_{pq}(a) = 0, q \neq p'.$$

Thus we obtain an $(n, G(\boldsymbol{A}))$-automaton $\boldsymbol{A}' = (\widehat{G(\boldsymbol{A})}_n, X, \delta_\Psi)$.

Now, we will prove that $\boldsymbol{A} \approx \boldsymbol{A}'$. To this end, we define the mapping ρ as follows:

> $\rho(s) = (he_{pi})$ for $s \in S$ where e is the identity of $G(\boldsymbol{A})$, i and h are, respectively, a number and an element in $G(\boldsymbol{A})$ uniquely determined such that $s_i \in T$ and $s = h(s_i)$.

Then $\rho(s) \in \widehat{G(\boldsymbol{A})}_n$ and ρ is a surjective mapping of S onto $\widehat{G(\boldsymbol{A})}_n$. Now, we will prove that for any $s \in S$ and any $a \in X$ we have $\rho(\delta(s, a)) = \delta_\Psi(\rho(s), a)$. Let $s = h(s_i)$ where $h \in G(\boldsymbol{A})$. Then we have $\rho(s) = (he_{pi})$. On the other hand, we have the following equality:

$$\delta_\Psi(\rho(s), a) = \rho(s)\Psi(a) = (he_{pi})(\psi_{pq}(a)) = \left(\sum_{k=1}^{n} he_{ki}\psi_{kp}(a)\right)$$
$$= (h\psi_{ip}(a)).$$

Assume that $\psi_{ip'}(a) \in G(\boldsymbol{A})$ and $\psi_{ip}(a) = 0, p \neq p'$. Then by the definition of $\Psi(a)$, we have $\delta(s_i, a) = \psi_{ip'}(a)(s_{p'})$. Therefore, we have $\delta(s, a) = \delta(h(s_i), a) = h(\delta(s_i, a)) = h(\psi_{ip'}(a)(s_{p'})) = (h\psi_{ip'}(a))(s_{p'})$. Cosequently, if we put $\rho(\delta(s, a)) = (\eta_p)$, then $\eta_{p'} = h\psi_{ip'}(a)$ and $\eta_p = 0, p \neq p'$ hold because of the definition of ρ. Thus we have $\delta_\Psi(\rho(s), a) = \rho(\delta(s, a))$. Therefore, $\boldsymbol{A} \approx \boldsymbol{A}'$ holds and also $\boldsymbol{A}'$ is strongly connected. Moreover, by Proposition 1.1.6, $G(\boldsymbol{A}) \approx G(\boldsymbol{A}')$ holds.

This means that $\boldsymbol{A}'$ is regular.

Corollary 1.3.1 *Let $\boldsymbol{A} = (S, X, \delta)$ be a perfect automaton and let G be a group such that $G \approx G(\boldsymbol{A})$. Then $\boldsymbol{A}$ is isomorphic to some $(1, G)$-automaton.*

Corollary 1.3.2 (the same as Proposition 1.1.5) *Let* $\boldsymbol{A} = (S, X, \delta)$ *be a strongly connected automaton. Then if* $|S| = |G(\boldsymbol{A})|$, $\boldsymbol{A}$ *is a permutation automaton.*

Proof By $|S| = |G(\boldsymbol{A})|$, $\boldsymbol{A}$ is isomorphic to some $(1, G)$-automaton. Then by Proposition 1.2.2, $\boldsymbol{A}$ is a permutation automaton.

Remark 1.3.1 Let $\boldsymbol{A} = (\widehat{G_n}, X, \delta_\Psi)$ be a regular (n, G)-automaton. Then $|\widehat{G_n}| = n|G|$ and $G(\boldsymbol{A}) \approx G$ hold.

Consequently, we have the following result:

The determination of any distinct strongly connected automata whose automorphism groups are isomorphic to a given finite group G is equivalent to that of any distinct regular group-matrix type automata of each positive integer's order on G.

Notice that we do not consider two isomorphic automata as distinct ones. Thus the following problem will be induced:

What conditions are required in order that two group-matrix type automata may be isomorphic to each other?

This problem will be treated in the following section.

Example 1.3.1 Figure 1.3 denotes the state transition diagram of a strongly connected automaton $\boldsymbol{A} = (S, X, \delta)$. We will obtain a representation of $\boldsymbol{A}$ by a group-matrix type automaton of order $|S|/|G(\boldsymbol{A})|$ on $G(\boldsymbol{A})$.

$G(\boldsymbol{A}) = \{e, \rho\}$ where e is the identity of $G(\boldsymbol{A})$ and $\rho = (p\ q)(t\ r)$, i.e. $\rho(p) = q$, $\rho(q) = p$, $\rho(t) = r$ and $\rho(r) = t$. $|S|/|G(\boldsymbol{A})| = 2$, $X = \{a, b, c\}$, $S = \{p, q, r, t\}$, $S_1 = \{p, q\}$, $S_2 = \{t, r\}$, $T = \{p, t\}$, $\delta(p, a) = q = \rho(p)$, $\delta(t, a) = t = e(t)$, $\Psi(a) = \begin{pmatrix} \rho & 0 \\ 0 & e \end{pmatrix}$, $\delta(p, b) = t = e(t)$, $\delta(t, b) = p = e(p)$, $\Psi(b) = \begin{pmatrix} 0 & e \\ e & 0 \end{pmatrix}$, $\delta(p, c) = p = e(p)$, $\delta(t, c) = r = \rho(t)$, $\Psi(c) = \begin{pmatrix} e & 0 \\ 0 & \rho \end{pmatrix}$.

Then $\boldsymbol{A}' = (\widehat{G(\boldsymbol{A})}_2, X, \delta_\Psi) \approx \boldsymbol{A}$.

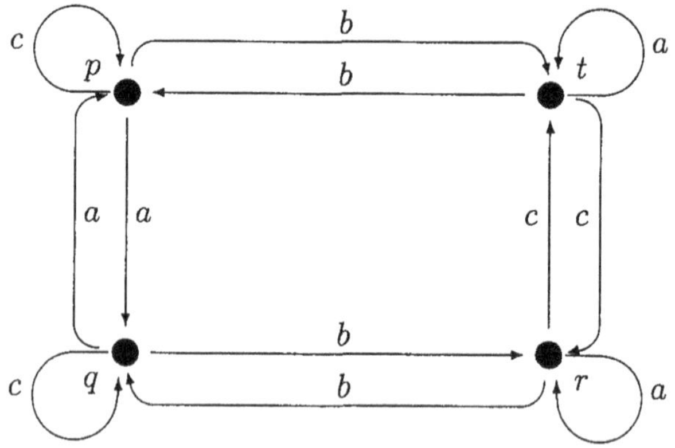

Figure 1.3: State transition diagram of $\boldsymbol{A}$

1.4 Equivalence of regular systems

In this section, we will deal with the problem as noted in the previous page.

Definition 1.4.1 Let G be a finite group, let n be a positive integer and let E be a subset of $\widetilde{G_n}$. Then E is called a *regular system* in $\widetilde{G_n}$ if there exists some regular (n, G)-automaton $\boldsymbol{A} = (\widehat{G_n}, X, \delta_\Psi)$ such that $E = \Psi(X)$.

Definition 1.4.2 Let E and F be two regular systems in $\widetilde{G_n}$. Then we say that E and F are *equivalent to each other* and denote $E \sim F$ if
there exist some permutation Θ on $\widehat{G_n}$ and isomorphism Φ of E^* onto F^* such that the following two conditions are satisfied:

(a) $F = \Phi(E)$.

(b) $\Theta(\hat{g}Y) = \Theta(\hat{g})\Phi(Y)$ for any $\hat{g} \in \widehat{G_n}$ and any $Y \in E$.

Here, $K^*, K \subseteq \widetilde{G_n}$, denotes the monoid generated by the elements of $K \cup \{(e_{pq})\}$ where e is the identity of G. Notice that, in this case,

$\sim$ induces an equivalence relation on the set of all regular systems in $\widetilde{G_n}$.

Before considering the relationship between regular systems and regular (n, G)-automata, we introduce the concept of the following generalized isomorphism.

Definition 1.4.3 Let $\boldsymbol{A} = (S, X, \delta)$ and $\boldsymbol{B} = (T, X, \gamma)$ be automata. Then $\boldsymbol{A}$ is said to be *isomorphic to* $\boldsymbol{B}$ *in the wider sense*, denoted by $\boldsymbol{A} \approx_w \boldsymbol{B}$, if there exist a bijective mapping ρ of S onto T and a permutation ξ on X such that $\rho(\delta(s, a)) = \gamma(\rho(s), \xi(a))$ for any $s \in S$ and $a \in X$.

Remark 1.4.1 It can be easily seen that $\boldsymbol{B} \approx_w \boldsymbol{A}$ if $\boldsymbol{A} \approx_w \boldsymbol{B}$.

Proposition 1.4.1 *Let* $\boldsymbol{A} = (S, X, \delta)$ *and* $\boldsymbol{B} = (T, X, \gamma)$ *be automata. If* $\boldsymbol{A} \approx_w \boldsymbol{B}$*, then* $G(\boldsymbol{A}) \approx G(\boldsymbol{B})$

Proof Since $\boldsymbol{A} \approx_w \boldsymbol{B}$, there exist a bijective mapping ρ of S onto T and a permutation ξ on X such that $\rho(\delta(s, a)) = \gamma(\rho(s), \xi(a))$ for any $s \in S$ and $a \in X$. Let $g \in G(\boldsymbol{A})$, let $t \in T$ and let $b \in X$. Consider the mapping $\rho g \rho^{-1}$. Since $t \in T$ and $b \in X$, there exist $s \in S$ and $a \in X$ such that $t = \rho(s)$ and $b \in \xi(a)$. Then $\rho g \rho^{-1}(\gamma(t, b)) = \rho g \rho^{-1}(\gamma(\rho(s), \xi(a)) = \rho g \rho^{-1}(\rho(\delta(s, a))) = \rho g(\delta(s, a)) = \rho(g(\delta(s, a))) = \rho(\delta(g(s), a)) = \gamma(\rho(g(s)), \xi(a)) = \gamma(\rho g \rho^{-1}(t), b)$. Hence $\rho g \rho^{-1} \in G(\boldsymbol{B})$. Since the correspondence $g \leftrightarrow \rho g \rho^{-1}$ is one-to-one and $gg' \leftrightarrow \rho g g' \rho^{-1} = (\rho g \rho^{-1})(\rho g' \rho^{-1})$ for any $g, g' \in G(\boldsymbol{A})$, we have $G(\boldsymbol{A}) \approx G(\boldsymbol{B})$.

Now we consider the relationship between regular systems and (n, G)-automata. In this case, it is enough to consider only simplified regular (n, G)-automata.

Theorem 1.4.1 *Let* $\boldsymbol{A} = (\widehat{G_n}, X, \delta_\Psi)$ *and* $\boldsymbol{A}' = (\widehat{G_m}, X, \delta_{\Psi'})$ *be simplified regular* (n, G)*-, and* (m, G)*-automaton, respectively. Then* $\boldsymbol{A} \approx_w \boldsymbol{A}'$ *if and only if two regular systems* $\Psi(X)$ *and* $\Psi'(X)$ *are equivalent.*

Proof Assume that $\boldsymbol{A} \approx_w \boldsymbol{A}'$. Then $n = m$ holds because of $|\widehat{G_n}| = |\widehat{G_m}|$. Therefore, $\Psi(X)$ and $\Psi'(X)$ are regular systems in the same $\widehat{G_n}$. Moreover, since $\boldsymbol{A} \approx_w \boldsymbol{A}'$, there exist a permutation ρ on $\widehat{G_n}$ and a permutation ξ on X such that $\rho(\delta_\Psi(\hat{g}, a)) = \delta_{\Psi'}(\rho(\hat{g}), \xi(a))$ for any $\hat{g} \in \widehat{G_n}$ and any $a \in X$. By this fact, we can verify easily that $\rho(\hat{g}\Psi(a_1)\Psi(a_2)\cdots\Psi(a_l)) = \rho(\hat{g})\Psi'(\xi(a_1))\Psi'(\xi(a_2))\cdots\Psi'(\xi(a_l))$ holds for any $a_i \in X, i = 1, 2, \ldots, l, l \geq 1$.

Now we consider the following mapping:

(1) $\Phi(\Psi(a)) = \Psi'(\xi(a))$ for any $a \in X$.

(2) $\Phi(\Psi(a_1)\Psi(a_2)\cdots\Psi(a_l)) = \Psi'(\xi(a_1))\Psi'(\xi(a_2))\cdots\Psi'(\xi(a_l))$ for any $a_i \in X, i = 1, 2, \ldots, l, l \geq 1$.

(3) $\Phi((e_{pq})) = (e_{pq})$.

First, we show that the above Φ can be well defined as a mapping. That is, we must show that $\Psi'(\xi(a_1))\Psi'(\xi(a_2))\cdots\Psi'(\xi(a_l)) = \Psi'(\xi(b_1))\Psi'(\xi(b_2))\cdots\Psi'(\xi(b_r))$ if $\Psi(a_1)\Psi(a_2)\cdots\Psi(a_l) = \Psi(b_1)\Psi(b_2)\cdots\Psi(b_r)$ where $a_i, b_j \in X, i = 1, 2, \ldots, l, j = 1, 2, \ldots, r, l, r \geq 1$, holds and that $\Psi'(\xi(a_1))\Psi'(\xi(a_2))\cdots\Psi'(\xi(a_l)) = (e_{pq})$ if $\Psi(a_1)\Psi(a_2)\cdots\Psi(a_l) = (e_{pq})$, $a_i \in X, i = 1, 2, \ldots, l, l \geq 1$ holds.

For this purpose, we assume that $\Psi(a_1)\Psi(a_2)\cdots\Psi(a_l) = \Psi(b_1)\Psi(b_2)\cdots\Psi(b_r)$ where $a_i, b_j \in X, i = 1, 2, \ldots, l, j = 1, 2, \ldots, r, l, r \geq 1$ holds. Then we have $\hat{g}\Psi(a_1)\Psi(a_2)\cdots\Psi(a_l) = \hat{g}\Psi(b_1)\Psi(b_2)\cdots\Psi(b_r)$ for any $\hat{g} \in \widehat{G_n}$. By the application of the mapping ρ to the above equality, we have $\rho(\hat{g})\Psi'(\xi(a_1))\Psi'(\xi(a_2))\cdots\Psi'(\xi(a_l)) = \rho(\hat{g})\Psi'(\xi(b_1))\Psi'(\xi(b_2))\cdots\Psi'(\xi(b_r))$ for any $\hat{g} \in \widehat{G_n}$. Since ρ is a surjective mapping of $\widehat{G_n}$ onto $\widehat{G_m}(= \widehat{G_n})$ and we can choose an arbitrary element $\hat{g} \in \widehat{G_n}$, $\Psi'(\xi(a_1))\Psi'(\xi(a_2))\cdots\Psi'(\xi(a_l)) = \Psi'(\xi(b_1))\Psi'(\xi(b_2))\cdots\Psi'(\xi(b_r))$ must hold.

Thus, we obtain the first part. As for the second part, we can prove it in a similar way. Consequently, Φ is well defined.

Next, we can prove easily that Φ is a surjective mapping of $\Psi(X)^*(= \Psi(X^*))$ onto $\Psi'(X)^*(= \Psi'(X^*))$. Moreover, it is easily seen that Φ is a homomorphism. Thus Φ is an isomorphism of $\Psi(X)^*$ onto $\Psi'(X)^*$ such that $\Psi'(X) = \Phi(\Psi(X))$. Therefore, we have the condition (a) in Definition 1.4.2.

Now, put $\Theta = \rho$. Moreover, notice that $\Phi(\Psi(a)) = \Psi'(\xi(a))$ holds for any $a \in X$. Then we have $\Theta(\hat{g}\Psi(a)) = \rho(\hat{g}\Psi(a)) = \rho(\delta_\Psi(\hat{g}, a))$. On the other hand, $\Theta(\hat{g})\Phi(\Psi(a)) = \rho(\hat{g})\Psi'(\xi(a)) = \delta_{\Psi'}(\rho(\hat{g}), \xi(a))$ holds. Since $\boldsymbol{A} \approx_w \boldsymbol{A}'$, we have $\Theta(\hat{g}\Psi(a)) = \Theta(\hat{g})\Phi(\Psi(a))$ for any $\hat{g} \in \widehat{G_n}$. This means that the condition (b) is satisfied.

Thus $\Psi(X)$ and $\Psi'(X)$ are equivalent.

Conversely, assume that $\Psi(X)$ and $\Psi'(X)$ are equivalent. Then, obviously we have $n = m$. Remember that Ψ and Ψ' are one-to-one because of Proposition 1.2.1. Moreover, let Φ be an isomorphism of $\Psi(X)^*$ onto $\Psi'(X)^*$ which satisfies the condition (a).

Now, consider the mapping ξ which is defined as follows:

$$\Psi'(\xi(a)) = \Phi(\Psi(a)) \text{ for any } a \in X.$$

In this case, we can see that $\xi(a)$ is determined uniquely for each $a \in X$, from the fact that Φ satisfies the condition (a) and that Ψ, Ψ' are surjective mappings. We can also see that ξ is a permutation on X.

Next, we put $\rho = \Theta$. Obviously, ρ is a surjective mapping of $\widehat{G_n}$ onto $\widehat{G_m}$. Then we have the following result:

> For any $\hat{g} \in \widehat{G_n}$ and any $a \in X$, $\rho(\delta_\Psi(\hat{g}, a)) = \rho(\hat{g}\Psi(a)) = \Theta(\hat{g}\Psi(a)) = \Theta(\hat{g})\Phi(\Psi(a)) = \rho(\hat{g})\Psi'(\xi(a)) = \delta_{\Psi'}(\rho(\hat{g}), \xi(a))$ holds.

Therefore, we have $\boldsymbol{A} \approx_w \boldsymbol{A}'$.

By Proposition 1.4.1, we have:

Corollary 1.4.1 *The set of all simplified regular (n, G)-automata equals $\{\boldsymbol{A} = (\widehat{G_n}, X, \delta_\Psi) \mid \Psi(X) :$ a regular system in $\widehat{G_n}\}$*

Remark 1.4.2 Notice that, in the case $n = 1$, the condition (a) implies the condition (b) in Definition 1.4.2.

Thus we have the following:

Corollary 1.4.2 *Let $\boldsymbol{A} = (\widehat{G_1}, X, \delta_\Psi)$ and $\boldsymbol{A}' = (\widehat{G_1}, X, \delta_{\Psi'})$ be simplified regular $(1, G)$-automata. Then $\boldsymbol{A} \approx_w \boldsymbol{A}'$ if and only if there exists an automorphism Φ of G such that $\Psi'(X) = \Phi(\Psi(X))$.*

Example 1.4.1 Let $S(3)$ be the symmetric group on $\{1,2,3\}$. Put $E = \{(1\ 2),(1\ 3)\}$ and $F = \{(1\ 2),(1\ 2\ 3)\}$. Then E and F are not equivalent though either is a regular system in $\widetilde{S(3)}_1$.

The condition in order that two regular systems may be equivalent is not so concrete except for the case $n = 1$. However, even in the general case, we can provide a criterion for the equivalence of regular systems in a concrete form.

First, we provide the following lemma:

Lemma 1.4.1 *Let E and F be two equivalent regular systems in $\widetilde{G_n}$, i.e. $E \sim F$. Moreover, assume that Θ and Φ are a permutation on $\widehat{G_n}$ and an isomorphism of E^* onto F^*, respectively, which satisfy the condition (a) and the condition (b) in Definition 1.4.2. Then there exist a permutation τ on $\{1,2,3,\ldots,n\}$ and n permutations $\Theta_i, i = 1,2,\ldots,n$ on G such that for any $\hat{g} = (g_1,g_2,\ldots,g_n) \in \widehat{G_n}$ we have $\Theta(\hat{g}) = (\widetilde{\Theta_1}(g_{\tau(1)}), \widetilde{\Theta_2}(g_{\tau(2)}),\ldots,\widetilde{\Theta_n}(g_{\tau(n)}))$ where each $\widetilde{\Theta_i}, i = 1,2,\ldots,n$, is the extension of Θ_i on G^0 with $\widetilde{\Theta_i}(0) = 0$.*

Proof Assume that the statement in the lemma is not true. Then there exist some elements $g,g',h,h' \in G, g \neq h$ and numbers $i,j,k = 1,2,\ldots,n, j \neq k$ such that $\Theta((ge_{pi})) = (g'e_{pj})$ and $\Theta((he_{pi})) = (h'e_{pk})$. On the other hand, by Definition 1.4.1 and by the proof of Theorem 1.4.1, there exist two regular (n,G)-automata $\boldsymbol{A} = (\widehat{G_n}, X, \delta_\Psi)$, $\boldsymbol{A}' = (\widehat{G_n}, X, \delta_{\Psi'})$ and a permutation ξ on X such that $\Psi(X) = E$, $\Psi'(X) = F$ and $\Theta(\delta_\Psi(\hat{g},a)) = \delta_{\Psi'}(\Theta(\hat{g}),\xi(a))$ for any $\hat{g} \in \widehat{G_n}$ and any $a \in X$. Then by the proof of Proposition 1.4.1, for any $\eta \in G(\boldsymbol{A})$ we have $\Theta\eta\Theta^{-1} \in G(\boldsymbol{A}')$. Now, put $\eta = \rho_{hg^{-1}} \in G(\boldsymbol{A})$ and compute the value $\Theta\eta\Theta^{-1}(g'e_{pj})$. We have $\Theta\eta\Theta^{-1}(g'e_{pj}) = \Theta\eta((ge_{pi})) = \Theta\rho_{hg^{-1}}((ge_{pi})) = \Theta((he_{pi})) = (h'e_{pk})$. Then $\Theta\eta\Theta^{-1} \in G(\boldsymbol{A}')$ is not represented by the form $\Theta\eta\Theta^{-1} = \rho_g$ for any $g \in G$. This contradicts the fact that $\boldsymbol{A}'$ is regular. Thus the assertion of the lemma must be true.

Making use of the above lemma, we can obtain the following result:

Theorem 1.4.2 *Let E and F be regular systems in $\widetilde{G_n}$. Then E and F are equivalent if and only if there exist n elements $k_i, i = 1, 2, \ldots, n$ in G, an automorphism φ of G and a permutation τ on $\{1, 2, 3, \ldots, n\}$ such that $F = \{(k_p^{-1}\tilde{\varphi}(x_{\tau(p)\tau(q)})k_q) \mid (x_{pq}) \in E\}$ where $\tilde{\varphi}$ is the extension of φ on G^0 with $\tilde{\varphi}(0) = 0$ and $k_1 = e$.*

Proof ($\Rightarrow$) Let Θ and Φ be a permutation on $\widehat{G_n}$ and an isomorphism of E^* onto F^*, respectively, which satisfy the condition (a) and the condition (b) in Definition 1.4.2. Then by Lemma 1.4.1, there exist a permutation τ on $\{1, 2, 3, \ldots, n\}$ and n permutations $\Theta_i, i = 1, 2, \ldots, n$ on G such that for any $\hat{g} = (g_1, g_2, \ldots, g_n) \in \widehat{G_n}$ we have $\Theta(\hat{g}) = (\widetilde{\Theta_1}(g_{\tau(1)}), \widetilde{\Theta_2}(g_{\tau(2)}), \ldots, \widetilde{\Theta_n}(g_{\tau(n)}))$ where each $\widetilde{\Theta_i}, i = 1, 2, \ldots, n$ is the extension of Θ_i on G^0 with $\widetilde{\Theta_i}(0) = 0$.

Now, put $Y = (y_{pq}) \in E^*$ and $\Phi(Y) = (z_{pq}) \in F^*$. Notice that $\hat{g}Y = (g_1, g_2, \ldots, g_n)(y_{pq}) = \left(\sum_{k=1}^{n} g_k y_{k1}, \sum_{k=1}^{n} g_k y_{k2}, \ldots, \sum_{k=1}^{n} g_k y_{kn}\right)$ holds for any $\hat{g} = (g_1, g_2, \ldots, g_n) \in \widehat{G_n}$.

Applying the mapping Θ to the above equality, we have:

$$\begin{aligned}\Theta((g_1, g_2, \ldots, g_n))\Phi((y_{pq}))\\ = \Theta\left(\left(\sum_{k=1}^{n} g_k y_{k1}, \sum_{k=1}^{n} g_k y_{k2}, \ldots, \sum_{k=1}^{n} g_k y_{kn}\right)\right).\end{aligned}$$

Thus we have $(\widetilde{\Theta_1}(g_{\tau(1)}), \widetilde{\Theta_2}(g_{\tau(2)}), \ldots, \widetilde{\Theta_n}(g_{\tau(n)}))(z_{pq})$

$$= \left(\widetilde{\Theta_1}\left(\sum_{k=1}^{n} g_k y_{k\tau(1)}\right), \widetilde{\Theta_2}\left(\sum_{k=1}^{n} g_k y_{k\tau(2)}\right), \ldots, \widetilde{\Theta_n}\left(\sum_{k=1}^{n} g_k y_{k\tau(n)}\right)\right).$$

The above equality implies that

$$\widetilde{\Theta_j}\left(\sum_{k=1}^{n} g_k y_{k\tau(j)}\right) = \sum_{k=1}^{n} \widetilde{\Theta_k}(g_{\tau(k)}) z_{kj}$$

for any $j = 1, 2, \ldots, n$.

Notice that

$$\widetilde{\Theta_j}\left(\sum_{k=1}^{n} g_k y_{k\tau(j)}\right) = \widetilde{\Theta_j}\left(\sum_{k=1}^{n} g_{\tau(k)} y_{\tau(k)\tau(j)}\right)$$

holds for any $j = 1, 2, \ldots, n$.

By the fact that E is a regular system in $\widetilde{G_n}$, for any $i = 1, 2, \ldots, n$ we can choose $g_{\tau(i)} = e$. Then we have $g_{\tau(u)} = 0$ for any $u, u \neq i$. From this result, we obtain $\widetilde{\Theta_i}(e)z_{ij} = \widetilde{\Theta_j}(y_{\tau(i)\tau(j)})$ for any $i, j = 1, 2, \ldots, n$. That is, $z_{ij} = \Theta_i(e)^{-1}\widetilde{\Theta_j}(y_{\tau(i)\tau(j)})$ holds for any $i, j = 1, 2, \ldots, n$.

Thus we can conclude as follows:

$\Phi((y_{pq})) = (h_p^{-1}\widetilde{\Theta_q}(y_{\tau(p)\tau(q)}))$ for any $(y_{pq}) \in E^*$ where $h_i = \Theta_i(e) \in G, i = 1, 2, \ldots, n$.

Let $(y_{pq}), (y'_{pq}) \in E^*$. Then $(y_{pq})(y'_{pq}) = \left(\sum_{k=1}^{n} y_{pk}y'_{kq}\right) \in E^*$. Since Φ is an isomorphism of E^* onto F^*, we obtain $\Phi((y_{pq}))\Phi((y'_{pq}))$ $= \Phi\left(\left(\sum_{k=1}^{n} y_{pk}y'_{kq}\right)\right)$.

Hence we have

$$(h_p^{-1}\widetilde{\Theta_q}(y_{\tau(p)\tau(q)}))(h_p^{-1}\widetilde{\Theta_q}(y'_{\tau(p)\tau(q)})) = \left(h_p^{-1}\widetilde{\Theta_q}\left(\sum_{k=1}^{n} y_{\tau(p)k}y'_{k\tau(q)}\right)\right)$$

and we have $\sum_{k=1}^{n}\widetilde{\Theta_k}(y_{\tau(i)\tau(k)})h_k^{-1}\widetilde{\Theta_j}(y'_{\tau(k)\tau(j)}) = \sum_{k=1}^{n}\widetilde{\Theta_j}(y_{\tau(i)k}y'_{k\tau(j)})$ for any $i, j = 1, 2, \ldots, n$. By the fact that E is a regular system in $\widetilde{G_n}$, we can see that for any $g \in G$ there exist some group-matrices (y_{pq}), $(y'_{pq}) \in E^*$ such that $y_{\tau(i)\tau(i)} = g$ and $y'_{\tau(i)\tau(j)} = e$. Thus we have $\Theta_i(g)h_i^{-1}h_j = \Theta_j(g)$ for any $g \in G$ and $i, j = 1, 2, \ldots, n$.

Let $\Theta((e, 0, \ldots, 0)) = (0, \ldots, h_t, 0, \ldots, 0), t = 1, 2, \ldots, n$. Put $\varphi(g) = h_t^{-1}\Theta_t(g), g \in G$ and $k_i = h_t^{-1}h_i \in G, i = 1, 2, \ldots, n$. Then we have $h_i^{-1}\widetilde{\Theta_j}(g) = h_i^{-1}\widetilde{\Theta_t}(g)h_t^{-1}h_j = k_i^{-1}\tilde{\varphi}(g)k_j$ for any $g \in G^0$ and $i, j = 1, 2, \ldots, n$.

First we prove that φ is an automorphism of G. To this end, it is enough to show that $\varphi(gg') = \varphi(g)\varphi(g')$ holds for any $g, g' \in G$. Since E is a regular system in $\widetilde{G_n}$, there exist $Y, Z \in E^*$ such that the $(1,1)$-entries of Y and Z are g and g', respectively. Then the $(1,1)$-entry of YZ becomes gg'. Notice that $(e, 0, \ldots, 0)Y = (g, 0, \ldots, 0), (e, 0, \ldots, 0)Z = (g', 0, \ldots, 0)$ and $(e, 0, \ldots, 0)YZ = (gg', 0, \ldots, 0)$. Therefore, we have $\Theta((g, 0, \ldots, 0)) = \Theta((e, 0, \ldots, 0)Y) = \Theta((e, 0, \ldots, 0))\Phi(Y) = (0, \ldots, 0, h_t, 0, \ldots, 0)\Phi(Y) = (0, \ldots, 0, \Theta_t(g),$

$0,\ldots,0), \Theta((g',0,\ldots,0)) = \Theta((e,0,\ldots,0)Z) = \Theta((e,0,\ldots,0))\Phi(Z) = (0,\ldots,0,h_t,0,\ldots,0)\Phi(Z) = (0,\ldots,0,\Theta_t(g'),0,\ldots,0)$ and $\Theta((gg', 0,\ldots,0)) = \Theta((e,0,\ldots,0)YZ) = \Theta((e,0,\ldots,0))\Phi(YZ) = (0,\ldots,0, h_t,0,\ldots,0)\Phi(YZ) = (0,\ldots,0,\Theta_t(gg'),0,\ldots,0)$. Hence the (t,t)-entries of $\Phi(Y), \Phi(Z)$ and $\Phi(YZ)$ are $h_t^{-1}\Theta_t(g), h_t^{-1}\Theta_t(g')$ and $h_t^{-1}\Theta_t (gg')$, respectively. Notice that $\Phi(YZ) = \Phi(Y)\Phi(Z)$. Consequently, $h_t^{-1}\Theta_t(gg') = h_t^{-1}\Theta_t(g)h_t^{-1}\Theta_t(g')$ and $\varphi(gg') = \varphi(g)\varphi(g')$.

Furthermore, we have $F = \Phi(E) = \{\Phi((y_{pq})) \mid (y_{pq}) \in E\} = \{(h_p^{-1}\widetilde{\Theta}_q(y_{\tau(p)\tau(q)})) \mid (y_{pq}) \in E\} = \{(k_p^{-1}\tilde{\varphi}(y_{\tau(p)\tau(q)})k_q) \mid (y_{pq}) \in E\}$.

($\Leftarrow$) Assume that there exist φ, $k_i, i = 1,2,\ldots,n$ and τ in the theorem. Let $\Theta((g_1,g_2,\ldots,g_n)) = (\tilde{\varphi}(g_{\tau(1)})k_1, \tilde{\varphi}(g_{\tau(2)})k_2,\ldots,\tilde{\varphi}(g_{\tau(n)}) k_n), (g_1,g_2,\ldots,g_n) \in \widehat{G_n}$. Then Θ is a permutation on $\widehat{G_n}$.

Let $\Phi(Y) = (k_p^{-1}\tilde{\varphi}(y_{\tau(p)\tau(q)})k_q)$ for any $Y = (y_{pq}) \in E$ and let $\Phi((e_{pq})) = (e_{pq})$. Moreover, let $Y = (y_{pq}) \in E$ and let $Z = (z_{pq}) \in E$.

Then we have $\Phi(Y)\Phi(Z) = (\sum_{i=1}^{n} k_p^{-1}\tilde{\varphi}(y_{\tau(p)\tau(i)})\tilde{\varphi}(z_{\tau(i)\tau(q)})k_q) = (\sum_{i=1}^{n} k_p^{-1}\tilde{\varphi}(y_{\tau(p)i}z_{i\tau(q)})k_q) = (k_p^{-1}\tilde{\varphi}(\sum_{i=1}^{n} y_{\tau(p)i}z_{i\tau(q)})k_q) = \Phi(YZ)$.

Hence Φ is an isomorphism of E^* onto F^*.

Now we prove that $\Theta(\hat{g}Y) = \Theta(\hat{g})\Phi(Y)$ for any $\hat{g} \in \widehat{G_n}$ and any $Y \in E$.

Let $\hat{g} = (g_1,g_2,\ldots,g_n) \in \widehat{G_n}$ and let $Y = (y_{pq}) \in E$.

Then we have $\Theta(\hat{g}Y) = \Theta((\sum_{t=1}^{n} g_t y_{t1}, \sum_{t=1}^{n} g_t y_{t2},\ldots,\sum_{t=1}^{n} g_t y_{tn}))$

$= (\tilde{\varphi}(\sum_{t=1}^{n} g_t y_{t\tau(1)})k_1, \tilde{\varphi}(\sum_{t=1}^{n} g_t y_{t\tau(2)})k_2,\ldots,\sum_{t=1}^{n} g_t y_{t\tau(n)})k_n)$

$= (\sum_{t=1}^{n}\tilde{\varphi}(g_t)\tilde{\varphi}(y_{t\tau(1)})k_1, \sum_{t=1}^{n}\tilde{\varphi}(g_t)\tilde{\varphi}(y_{t\tau(2)})k_2,\ldots,\sum_{t=1}^{n}\tilde{\varphi}(g_t)\tilde{\varphi}(y_{t\tau(n)})k_n)$.

On the other hand, we have $\Theta(\hat{g})\Phi(Y)$

$= (\tilde{\varphi}(g_{\tau(1)})k_1, \tilde{\varphi}(g_{\tau(2)})k_2,\ldots,\tilde{\varphi}(g_{\tau(n)})k_n)(k_p^{-1}\tilde{\varphi}(y_{\tau(p)\tau(q)})k_q)$

$= (\sum_{t=1}^{n}\tilde{\varphi}(g_{\tau(t)})\tilde{\varphi}(y_{\tau(t)\tau(1)})k_1, \sum_{t=1}^{n}\tilde{\varphi}(g_{\tau(t)})\tilde{\varphi}(y_{\tau(t)\tau(2)})k_2,\ldots,$

$\sum_{t=1}^{n}\tilde{\varphi}(g_{\tau(t)})\tilde{\varphi}(y_{\tau(t)\tau(n)})k_n) = (\sum_{t=1}^{n}\tilde{\varphi}(g_t)\tilde{\varphi}(y_{t\tau(1)})k_1, \sum_{t=1}^{n}\tilde{\varphi}(g_t)\tilde{\varphi}(y_{t\tau(2)})k_2,$

$$\ldots, \sum_{t=1}^{n} \tilde{\varphi}(g_t)\tilde{\varphi}(y_{t\tau(n)})k_n).$$

Consequently, $\Theta(\hat{g}Y) = \Theta(\hat{g})\Phi(Y)$ holds and hence $E \sim F$.

1.5 Characteristic monoids and input sets

This section consists of two applications of our method. One is concerned with the characteristic monoids of automata, and the other is concerned with the input sets of automata.

Definition 1.5.1 Let $\boldsymbol{A} = (S, X, \delta)$ be an automaton and let $x, y \in X^*$. Then x and y are *equivalent to each other* and we denote $x \sim y$ if $\delta(s, x) = \delta(s, y)$ holds for any $s \in S$. We denote by $\overline{x}$ the set of all $y \in X^*$ such that $x \sim y$ and by $C(\boldsymbol{A})$ the set of all such classes, i.e. $C(\boldsymbol{A}) = \{\overline{x} \mid x \in X^*\}$. Notice that $C(\boldsymbol{A})$ forms a monoid under the natural operation, i.e. $\overline{x} \cdot \overline{y} = \overline{xy}$ where $x, y \in X^*$. This monoid is called the *characteristic monoid* of $\boldsymbol{A}$.

Proposition 1.5.1 *Let $\boldsymbol{A} = (S, X, \delta)$ and $\boldsymbol{B} = (T, X, \gamma)$ be two isomorphic automata. Then $C(\boldsymbol{A}) = C(\boldsymbol{B})$.*

Proof Let ρ be an isomorphism of $\boldsymbol{A}$ onto $\boldsymbol{B}$. Then for any $x, y \in X^*$ and $s \in S$, $\delta(s, x) = \delta(s, y)$ if and only if $\gamma(\rho(s), x) = \gamma(\rho(s), y)$. Since ρ is a bijective mapping, this means that, for any $s \in S$, $\delta(s, x) = \delta(s, y)$ if and only if $\gamma(s, x) = \gamma(s, y)$. Hence $C(\boldsymbol{A}) = C(\boldsymbol{B})$.

Proposition 1.5.2 *Let $\boldsymbol{A} = (\widehat{G_n}, X, \delta_\Psi)$ be an (n, G)-automaton. Then the characteristic monoid of $\boldsymbol{A}$ is isomorphic to $\Psi(X^*)$.*

Proof Obvious from the fact that, for any $x, y \in X^*$, $x \sim y$ if and only if $\Psi(x) = \Psi(y)$.

From the above proposition, to study the structure of the characteristic monoid of a strongly connected automaton, it is enough to study that of the corresponding regular group-matrix type automaton.

Thus the following result in [57] can be proved by our method.

Proposition 1.5.3 *Let $\boldsymbol{A} = (S, X, \delta)$ be a strongly connected automaton. Then $C(\boldsymbol{A})$ forms a group if and only if $\boldsymbol{A}$ is a permutation automaton.*

Proof By Theorem 1.3.1, we can assume that $\boldsymbol{A}$ is a regular group-matrix type automaton, i.e. $\boldsymbol{A} = (\widehat{G_n}, X, \delta_\Psi)$.

($\Leftarrow$) In this case, for any $Y \in \Psi(X^*)$ there exist a permutation τ on $\{1, 2, \ldots, n\}$ and elements $g_{ij} \in G, i, j = 1, 2, \ldots, n$ such that $Y = \left(g_{pq}e_{p\tau(q)}\right)$ where e is the identity of G.

Let m be a positive integer such that $\tau^m = 1$ where 1 is the identity permutation on $\{1, 2, \ldots, n\}$. Then there exist some elements $h_{ij} \in G, i, j = 1, 2, \ldots, n$ such that $Y^m = (h_{pq}e_{pq}) \in \Psi(X^*)$. Since G is a finite group, there exists a positive integer r such that $(Y^m)^r = (e_{pq})$. Notice that (e_{pq}) is the identity of $\Psi(X^*)$ and that $\Psi(X^*)$ is a finite monoid. Therefore, $\Psi(X^*)$ must be a group. By Proposition 1.5.2, $C(\boldsymbol{A})$ forms a group.

($\Rightarrow$) For any $Y \in \Psi(X^*)$, Y is of the form $\left(g_{pq}e_{p\tau(q)}\right)$ where $g_{ij} \in G, i, j = 1, 2, \ldots, n$ and τ is a permutation on $\{1, 2, \ldots, n\}$. Because, if it is not the case, $Y^m \neq (e_{pq})$ for any positive integer m.

This contradicts the assumption that $C(\boldsymbol{A})$ is a group. Thus the assertion is true and in this case, $\boldsymbol{A}$ is a permutation automaton.

Now, we establish a relationship between the input sets and the state sets of strongly connected automata.

Theorem 1.5.1 *Let $\boldsymbol{A} = (S, X, \delta)$ be a strongly connected automaton whose automorphism group is isomorphic to a finite group G. Then we have $|S||X| \geq I(G)|G|$ where $I(G) = min\{|H| \mid H \subseteq G, [H] = G\}$ and $[H]$ is the subgroup of G generated by the elements of H.*

Proof We can assume that $\boldsymbol{A}$ is of the form $\boldsymbol{A} = (\widehat{G_n}, X, \delta_\Psi)$, i.e. $\boldsymbol{A}$ is a regular (n, G)-automaton. Then it is enough to show that $n\,|G|\,|X| \geq I(G)\,|G|$.

Let $\Psi(a)^\sharp$ be the set of all nonzero components of $\Psi(a)$ where $a \in X$. Then obviously $\left|\Psi(a)^\sharp\right| \leq n$ holds. By the strong connectedness of $\boldsymbol{A}$, we obtain immediately $\left[\bigcup_{a \in X} \Psi(a)^\sharp\right] = G$. Thus we have

$\left|\bigcup_{a\in X}\Psi(a)^\sharp\right| \geq I(G)$. On the other hand, $n\,|X| \geq \left|\bigcup_{a\in X}\Psi(a)^\sharp\right|$ holds. Hence we have $n\,|G|\,|X| \geq I(G)\,|G|$.

From the above theorem, we can see that there is no strongly connected automaton $\boldsymbol{A} = (S, X, \delta)$ such that $|X| < I(G)/n$ where $n = |S|\,/\,|G(\boldsymbol{A})|$ and $G(\boldsymbol{A}) \approx G$.

Thus we may have the following question:

> *Can we construct an automaton with the smallest cardinality of input set among the strongly connected automata whose automorphism groups are isomorphic to a given finite group?*

In response to this question, for a finite group G and a positive integer n, we define the number $J(n, G)$ as follows: $J(n, G) = min\{|X| \mid \boldsymbol{A} = (S, X, \delta)$ is a strongly connected automaton such that $G(\boldsymbol{A}) \approx G$ and $|S| = n|G(\boldsymbol{A})|\}$.

Let r be a positive number. By $\lceil r \rceil$, we denote the positive integer m such that $m - 1 < r \leq m$. Then we have the following result.

Theorem 1.5.2 *Let G be a finite group and let n be a positive integer. Then $\lceil I(G)/n \rceil \leq J(n, G) \leq \lceil I(G)/n \rceil + p(n)$ where $p(1) = 0$ and $p(n) = 1$ for $n \geq 2$.*

Proof Obviously, the theorem holds true for the case $n = 1$. Therefore, we consider the case $n \geq 2$.

The inequality $\lceil I(G)/n \rceil \leq J(n, G)$ is immediate from Theorem 1.5.1. Hence we have to prove the inequality $J(n, G) \leq \lceil I(G)/n \rceil + p(n)$.

By the definition of $I(G)$, there exists a set of generators H of G, i.e. $[H] = G$ such that $H = \{h_i \mid h_i \in G, i = 1, 2, \ldots, I(G)\}$. Now, put $X = Y \cup \{z\}$ where $Y = \{y_i \mid i = 1, 2, \ldots, \lceil I(G)/n \rceil\}$. Moreover, for any $i = 1, 2, \ldots, \lceil I(G)/n \rceil$, we can define $\Psi(y_i) \in \widetilde{G_n}$ such that all elements of $\Psi(y_i)^\sharp$ are gathered only into the first column of $\Psi(y_i)$, $\Psi(y_i) \neq \Psi(y_j)$ $(i \neq j)$ and $H = \bigcup_{i=1}^{\lceil I(G)/n \rceil} \Psi(y_i)^\sharp$.

Put $\tau = (1\,2\,3\,\ldots\,n)$ be the element of the symmetric group $S(n)$ on $\{1, 2, 3, \ldots, n\}$. Moreover, we assign $\Psi(z) = \left(e_{p\tau(q)}\right) \in \widetilde{G_n}$ where e is the identity of G. Thus we can define an (n, G)-automaton $\boldsymbol{A} = (\widehat{G_n}, X, \delta_\Psi)$.

Now, we prove that $\boldsymbol{A}$ is regular.

Proof of the strong connectedness of $\boldsymbol{A}$ First, we prove that, for any $i, j = 1, 2, \ldots, n$ and all $h \in H$, there exists an element $x \in X^*$ such that the (i, j)-entry of $\Psi(x)$ is equal to h.

By the assignment of $\Psi(Y)$, for any $h \in H$ there exist some integers $s = 1, 2, \ldots, n$ and $t = 1, 2, \ldots, \lceil I(G)/n \rceil$ such that the $(s, 1)$-entry of $\Psi(y_t)$ equals h. Now, put $x = z^u y_t z^{n-j+1}$ where $u \equiv i - s \pmod n$ and $u > 0$. Then it can be seen that the (i, j)-entry of $\Psi(x)$ equals h.

From this fact and the finiteness of G, it follows that, for any $i, j = 1, 2, \ldots, n$ and any $g \in G$, there exists an element $x \in X^*$ such that the (i, j)-entry of $\Psi(x)$ is equal to g. By Theorem 1.2.2, this indicates the strong connectedness of $\boldsymbol{A}$.

Proof of the regularity of $\boldsymbol{A}$ Let y be an arbitrary element of Y, let g be the $(1, 1)$-entry of $\Psi(y)$ and let k be the order of g. Then the $(1, 1)$-entry of $\Psi(y^k)$ is equal to e. On the other hand, the $(1, 1)$-entry of $\Psi(z^n)$ is also equal to e. By a comparison of these two group-matrices and by Theorem 1.2.4 and 1.2.5, $\boldsymbol{A}$ is regular.

Thus the existence of an automaton $\boldsymbol{A} = (S, X, \delta)$ such that $G(\boldsymbol{A}) \approx G$, $|S| = n\,|G(\boldsymbol{A})|$ and $|X| = \lceil I(G)/n \rceil + 1$ has been shown. This completes the proof of the theorem.

Example 1.5.1 Let G be the Klein's four group whose generators are g and h, satisfying the defining relations $g^2 = e$, $h^2 = e$ and $ghg^{-1}h^{-1} = e$. Then we can obtain the following two-input regular $(2, G)$-automaton $\boldsymbol{A} = (\widehat{G_2}, X, \delta_\Psi)$. $G = \{e, g, h, gh\}$, $X = \{g, h\}$, $\Psi(a) = \begin{pmatrix} g & 0 \\ h & 0 \end{pmatrix}$, $\Psi(b) = \begin{pmatrix} 0 & e \\ e & 0 \end{pmatrix}$.

Now, we put $s_1 = (e, 0)$, $s_2 = (g, 0)$, $s_3 = (h, 0)$, $s_4 = (gh, 0)$, $s_5 = (0, e)$, $s_6 = (0, g)$, $s_7 = (0, h)$ and $s_8 = (0, gh)$. Then we obtain the following state transition diagram of $\boldsymbol{A}$:

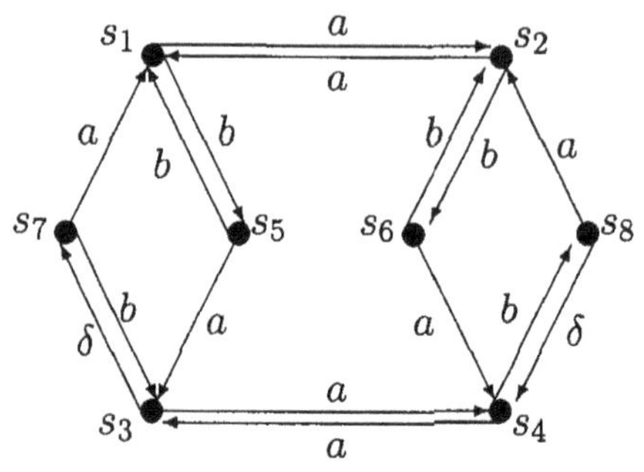

Figure 1.4: State transition diagram of $\boldsymbol{A}$

Theorem 1.5.3 *Let G be a finite group. Then there exists a strongly connected automaton $\boldsymbol{A} = (S, X, \delta)$ such that $G(\boldsymbol{A}) \approx G$ and $|X| \leq 2$.*

Proof To prove the theorem, it is enough to show that $min\{J(n,G) \mid n \geq 1\} \leq 2$. This is immediate from the fact that $J(n,G) \leq \lceil I(G)/n \rceil + p(n)$ and $min\{\lceil I(G)/n \rceil + p(n) \mid n \geq 1\} \leq 2$.

Remark 1.5.1 There is the case $J(n,G) = \lceil I(G)/n \rceil$. For instance, we consider the following case.

Let G be the symmetric group $[\{g,h,k\}] \subset S(9)$ where $g = (1\,2\,3)$, $h = (4\,5\,6)$ and $k = (7\,8\,9)$. Then obviously we have $I(G) = 3$ and thus $\lceil I(G)/2 \rceil = 2$. Now, put $X = \{a,b\}$, $\Psi(a) = \begin{pmatrix} g & 0 \\ h & 0 \end{pmatrix}$ and $\Psi(b) = \begin{pmatrix} 0 & k \\ k & 0 \end{pmatrix}$.

Then $\boldsymbol{A} = (\widehat{G_2}, X, \delta_\Psi)$ is a regular $(2,G)$-automaton. Notice here a role of b^3, i.e. $\Psi(b^3) = \begin{pmatrix} 0 & k^3 \\ k^3 & 0 \end{pmatrix} = \begin{pmatrix} 0 & e \\ e & 0 \end{pmatrix}$.

Thus we have $J(2,G) = \lceil I(G)/2 \rceil = 2$.

In the same way as the above, we can prove:

Theorem 1.5.4 *Let n and k be positive integers which are relatively prime, and let G be a finite group such that $I(G) \equiv 1 \pmod{n}$.*

Moreover, assume that there exist a set of generators H of G and an element $h \in H$ such that $|H| = I(G)$ and $o(h) \equiv k \pmod{n}$ where $o(h)$ means the order of h. Then we have $J(n, G) = \lceil I(G)/n \rceil$.

Corollary 1.5.1 *Let G be a finite group such that $I(G)\,|G|$ is an odd number. Then we have $J(2, G) = \lceil I(G)/2 \rceil$.*

Proof This is the case where we put, in the above theorem, $n = 2$, $k = 1$, $I(G) \equiv 1 \pmod{2}$ and $o(h) \equiv 1 \pmod{2}$. Here, h is an arbitrary element in H.

1.6 Direct product of automata

In this section, we will deal with direct products of group-matrix type automata and their automorphism groups.

Definition 1.6.1 Let $\boldsymbol{A} = (S, X, \delta)$ and $\boldsymbol{B} = (T, X, \gamma)$ be two automata. The *direct product*, $\boldsymbol{A} \times \boldsymbol{B}$, is the automaton $\boldsymbol{A} \times \boldsymbol{B} = (S \times T, X, \delta \times \gamma)$ where $\delta \times \gamma((s,t), a) = (\delta(s,a), \gamma(t,a))$ for any $(s,t) \in S \times T$ and $a \in X$.

Before introducing the notion of the direct product of group-matrix type automata, we define the product of two group-matrices.

Definition 1.6.2 Let H and K be subgroups of a finite group G, respectively. Moreover, assume that $Y = (y_{ij}), i, j = 1, 2, \ldots, n$ and $Z = (z_{pq}), p, q = 1, 2, \ldots, m$ are group-matrices of order n on H and of order m on K, respectively, i.e. $Y \in \widetilde{H_n}$ and $Z \in \widetilde{K_m}$. Then the *K-product (Kronecker product)* $Y \otimes Z$ of $Y = (y_{ij})$ and $Z = (z_{pq})$ is defined as follows:

$$Y \otimes Z = \begin{pmatrix} y_{11}Z & y_{12}Z & \cdots & y_{1n}Z \\ y_{21}Z & y_{22}Z & \cdots & y_{2n}Z \\ \vdots & \vdots & \ddots & \vdots \\ y_{n1}Z & y_{n2}Z & \cdots & y_{nn}Z \end{pmatrix}.$$

Notice that $Y \otimes Z$ is a group-matrix of order nm on G, i.e. $Y \otimes Z \in \widetilde{G_{nm}}$.

The following proposition can be easily proved.

Proposition 1.6.1 *Let H and K be subgroups of a finite group G such that $hk = kh$ for any $h \in H$ and any $k \in K$. Then we have $(Y_1 \otimes Z_1)(Y_2 \otimes Z_2) = (Y_1 Y_2) \otimes (Z_1 Z_2)$ for any $Y_1, Y_2 \in \widetilde{H_n}$ and $Z_1, Z_2 \in \widetilde{K_m}$.*

Definition 1.6.3 Let H and K be subgroups of a finite group G. Moreover, we assume that Ψ and Π are mappings of X, into $\widetilde{H_n}$ and into $\widetilde{K_m}$, respectively. Then the K-*product* $\Psi \otimes \Pi$ of Ψ and Π is defined as follows:

$$\Psi \otimes \Pi(a) = \Psi(a) \otimes \Pi(a) \text{ for any } a \in X.$$

Under the same notations as above, we have:

Proposition 1.6.2 *Assume that $hk = kh$ holds for any $h \in H$ and any $k \in K$. Then we have $\Psi \otimes \Pi(x) = \Psi(x) \otimes \Pi(x)$ for any $x \in X^*$. Here $\Psi \otimes \Pi$, Ψ and Π are extended, respectively, to the mappings of X^* in the natural way as indicated in Remark 1.2.1.*

Definition 1.6.4 Let H and K be subgroups of a finite group G. Moreover, we assume that $\boldsymbol{A} = (\widehat{H_n}, X, \delta_\Psi)$ and $\boldsymbol{B} = (\widehat{K_m}, X, \delta_\Pi)$ are an (n, H)-, and an (m, K)-automaton, respectively. Then the K-*product* $\boldsymbol{A} \otimes \boldsymbol{B}$ of $\boldsymbol{A}$ and $\boldsymbol{B}$ is the (nm, G)-automaton $\boldsymbol{A} \otimes \boldsymbol{B} = (\widehat{G_{nm}}, X, \delta_{\Psi \otimes \Pi})$.

In what follows, we will deal with the case $G = H \times K$.

Theorem 1.6.1 *Let G be the direct product of finite groups H and K, i.e. $G = H \times K$. Moreover, assume that $\boldsymbol{A} = (\widehat{H_n}, X, \delta_\Psi)$ and $\boldsymbol{B} = (\widehat{K_m}, X, \delta_\Pi)$ are an (n, H)-, and an (m, K)-automaton, respectively. Then we have $\boldsymbol{A} \otimes \boldsymbol{B} \approx \boldsymbol{A} \times \boldsymbol{B}$.*

Proof For any $\bar{f} = (h_1, h_2, \ldots, h_n, k_1, k_2, \ldots, k_m) \in \widehat{H_n} \times \widehat{K_m}$, we define the mapping ρ as follows:

$$\rho(\bar{f}) = (h_1 k_1, h_1 k_2, \ldots, h_1 k_m, h_2 k_1, h_2 k_2, \ldots, h_2 k_m, \ldots, h_n k_1,$$
$$h_n k_2, \ldots, h_n k_m).$$

Then $\rho(\bar{f}) \in \hat{G_{nm}}$ holds. Moreover, since $G = H \times K$, ρ is a bijective mapping of $\widehat{H_n} \times \widehat{K_m}$ onto $\widehat{G_{nm}}$.

Now, we prove that $\rho(\delta_\Psi \times \delta_\Pi(\bar{f}, a)) = \delta_{\Psi\otimes\Pi}(\rho(\bar{f}), a)$ holds for any $a \in X$ and any $\bar{f} \in \widehat{H_n} \times \widehat{K_m}$.

Let $\bar{f} = (h_1, h_2, \ldots, h_n, k_1, k_2, \ldots, k_m) \in \hat{H}_n \times \widehat{K_m}$, let $a \in X$ and let $\delta_\Psi \times \delta_\Pi(\bar{f}, a) = ((h_1, h_2, \ldots, h_n)\Psi(a), (k_1, k_2, \ldots, k_m)\Pi(a)) = (h'_1, h'_2, \ldots, h'_n, k'_1, k'_2, \ldots, k'_m)$. Then we have the following:

$$h'_i = \sum_{s=1}^{n} h_s \psi_{si}(a), i = 1, 2, \ldots, n, k'_j = \sum_{t=1}^{m} k_t \pi_{tj}(a), j = 1, 2, \ldots, m.$$

Here, we have put $\Psi(a) = (\psi_{pq}(a)), p, q = 1, 2, \ldots, n$ and $\Pi(a) = (\pi_{rs}(a)), r, s = 1, 2, \ldots, m$.

Hence the $[(i-1)m + j]$-th entry of $\rho(\delta_\Psi \times \delta_\Pi(\bar{f}, a)), \in \widehat{G_{nm}}$ is computed as follows:

$$h'_i k'_j = \left(\sum_s h_s \psi_{si}(a)\right)\left(\sum_t k_t \pi_{tj}(a)\right) = \sum_{s,t} h_s k_t \psi_{si}(a) \pi_{tj}(a).$$

On the other hand, we put $\delta_{\Psi\otimes\Pi}(\rho(\bar{f}), a) = \rho(\bar{f})\left(\Psi \otimes \Pi\right)(a) = (\gamma_1, \gamma_2, \ldots, \gamma_{nm})$.

Now, we compute the value $\gamma_{(i-1)m+j}$. First, notice that $h_s k_t, s = 1, 2, \ldots, n, t = 1, 2, \ldots, m$ is the $[(s-1)m + t]$-th entry of $\rho(\bar{f})$. Put here $\left(\Psi \otimes \Pi\right)(a) = (\xi_{\alpha\beta}), \alpha, \beta = 1, 2, \ldots, nm$. Thus we have $\gamma_{(i-1)m+j} = \sum_{s,t} h_s k_t \xi_{[(s-1)m+t][(i-1)m+j]}$. Since

$$\left(\Psi \otimes \Pi\right)(a) = \begin{pmatrix} \psi_{11}(a)\Pi(a) & \cdots & \psi_{1n}(a)\Pi(a) \\ \psi_{21}(a)\Pi(a) & \cdots & \psi_{2n}(a)\Pi(a) \\ \vdots & \cdots & \vdots \\ \psi_{n1}(a)\Pi(a) & \cdots & \psi_{nn}(a)\Pi(a) \end{pmatrix}$$

and $\Pi(a)$ is an $m \times m$ matrix, $\xi_{[(s-1)m+t][(i-1)m+j]} = \psi_{si}(a)\pi_{tj}(a)$ holds.

Consequently, we have $\gamma_{(i-1)m+j} = \sum_{s,t} h_s k_t \psi_{si}(a) \pi_{tj}(a)$. Therefore, we have $\rho(\delta_\Psi \times \delta_\Pi(\bar{f}, a)) = \delta_{\Psi\otimes\Pi}(\rho(\bar{f}), a)$ and thus $\boldsymbol{A} \otimes \boldsymbol{B} \approx \boldsymbol{A} \times \boldsymbol{B}$.

Corollary 1.6.1 *Let G be the direct product of finite groups H and K, i.e. $G = H \times K$. Moreover, we assume that $\boldsymbol{A} = (\widehat{H_n}, X, \delta_\Psi)$ and $\boldsymbol{B} = (\widehat{K_m}, X, \delta_\Pi)$ are an (n, H)-, and an (m, K)-automaton, respectively. Then $\boldsymbol{A} \times \boldsymbol{B}$ is a strongly connected automaton with $G(\boldsymbol{A} \times \boldsymbol{B}) \approx G$ if and only if $\{\Psi \otimes \Pi(a) \mid a \in X\}$ is a regular system in $\widetilde{G_{nm}}$. If, in addition, $\boldsymbol{A}$ and $\boldsymbol{B}$ are regular, then $G(\boldsymbol{A} \times \boldsymbol{B}) \approx G$ mentioned above can be replaced by $G(\boldsymbol{A} \times \boldsymbol{B}) \approx G(\boldsymbol{A}) \times G(\boldsymbol{B})$.*

1.7 Factor automata

In this section, we will consider the factor automata of group-matrix type automata. An application will appear in a decomposition theorem of automata.

Definition 1.7.1 Let $\boldsymbol{A} = (S, X, \delta)$ be an automaton and let $G(\boldsymbol{A})$ be its automorphism group. Moreover, assume that H is a subgroup of $G(\boldsymbol{A})$. Then the *factor automaton* $\boldsymbol{A}/H$ is the automaton $\boldsymbol{A}/H = (\overline{S_H}, X, \overline{\delta_H})$ where $\overline{S_H} = \{\overline{s} \mid s \in S\}$, $\overline{s} = \{t \mid t \in S$ and there exists $h \in H$ such that $t = h(s)\}$ and $\overline{\delta_H}(\overline{s}, a) = \overline{\delta(s, a)}$ for any $\overline{s} \in \overline{S_H}$ and any $a \in X$.

To investigate factor automata of group-matrix type automata, we define the following:

Definition 1.7.2 Let $\boldsymbol{A} = (\widehat{G_n}, X, \delta_\Psi)$ be an (n, G)-automaton and let ξ be a homomorphism of G onto some group. Then $\boldsymbol{A}_\xi$ is the $(n, \xi(G))$-automaton $\boldsymbol{A}_\xi = (\widehat{\xi(G)}_n, X, \delta_{\xi[\Psi]})$ such that $\xi[\Psi](a) = \left(\widetilde{\xi}(\psi_{pq}(a))\right)$ for any $a \in X$ where $\widetilde{\xi}$ is the extension of ξ on G^0 with $\widetilde{\xi}(0) = 0$, and $\Psi(a) = (\psi_{pq}(a))$.

The following result is immediate.

Proposition 1.7.1 *Let $\boldsymbol{A} = (\widehat{G_n}, X, \delta_\Psi)$ be a strongly connected (n, G)-automaton. Then $\boldsymbol{A}_\xi$ is also strongly connected.*

The following theorem plays a fundamental role in this section.

Theorem 1.7.1 *Let $\boldsymbol{A} = (\widehat{G_n}, X, \delta_\Psi)$ be a regular (n, G)-automaton and ξ be a homomorphism of G onto some group. Then we have $\boldsymbol{A}_\xi \approx \boldsymbol{A}/\,\mathrm{Ker}(\xi)$ where $\mathrm{Ker}(\xi)$ means the kernel of ξ, i.e. $\mathrm{Ker}(\xi) = \xi^{-1}(\xi(e))$ (e is the identity of G).*

Remark 1.7.1 In the above, we identify $G(\boldsymbol{A})$ with G. That is, we identify $\rho_g \in G(\boldsymbol{A})$ with $g \in G$ (for the notation, see the proof of Theorem 1.2.1). Consequently, $\mathrm{Ker}(\xi)$ is assumed to be a subgroup of $G(\boldsymbol{A})$. This assumption will be taken throughout this section.

Proof of Theorem 1.7.1 Put $H = Ker(\xi) \subseteq G$ and $\boldsymbol{A}/H = (\overline{S_H}, X, \overline{\delta_H})$. Then for $\widehat{f} = (f_p) \in \widehat{G_n}$, $\widehat{k} = (k_p) \in \widehat{G_n}$ we have the following:

$\overline{\widehat{f}} = \overline{\widehat{k}} \in \overline{S_H}$ holds if and only if there exists an element $h \in H$ such that $\widehat{f} = h(\widehat{k}) = \rho_h(\widehat{k}) = (hk_p)$.

Now, we define the mapping Ξ as follows:

$$\Xi(\overline{\hat{g}}) = \left(\widetilde{\xi}(g_p)\right) \quad \text{for any } \hat{g} = (g_p) \in \widehat{G_n}.$$

Then, Ξ is well defined as a mapping and it is a surjective mapping of $\overline{S_H}$ onto $\xi\widehat{(G)}_n$.

Now, we will show that $\boldsymbol{A}_\xi \approx \boldsymbol{A}/H$ holds. Since $\overline{\delta_H}(\overline{\hat{g}}, a) = \overline{\delta_\Psi(\hat{g}, a)} = \overline{\hat{g}\Psi(a)}$ holds for any $\hat{g} \in \widehat{G_n}$ and any $a \in X$, we have $\Xi(\overline{\delta_H}(\overline{\hat{g}}, a)) = \Xi(\overline{\hat{g}\Psi(a)})$. Now, we put $\hat{g} = (g_p) \in \widehat{G_n}$ and $\Psi(a) = (\psi_{pq}(a))$. Then we have $\hat{g}\Psi(a) = \left(\sum_{k=1}^{n} g_k\psi_{kp}(a)\right)$.

Thus we have $\Xi(\overline{\delta_H}(\overline{\hat{g}}, a)) = \Xi(\overline{\hat{g}\Psi(a)}) = \left(\widetilde{\xi}\left(\sum_{k=1}^{n} g_k\psi_{kp}(a)\right)\right) = \left(\sum_{k=1}^{n} \widetilde{\xi}(g_k)\widetilde{\xi}(\psi_{kp}(a))\right) = \left(\widetilde{\xi}(g_p)\right)\left(\widetilde{\xi}(\psi_{pq}(a))\right) = \Xi(\overline{\hat{g}})\xi[\Psi](a) = \delta_{\xi[\Psi]}(\Xi(\overline{\hat{g}}), a)$.

Therefore, we have $\boldsymbol{A}_\xi \approx \boldsymbol{A}/H$.

Corollary 1.7.1 *Let $\boldsymbol{A} = (\widehat{G_n}, X, \delta_\Psi)$ be a regular (n, G)-automaton and ξ be a homomorphism of G onto some group. Then $\xi(G)$ is isomorphic to a subgroup of $G(\boldsymbol{A}/\,\mathrm{Ker}(\xi))$.*

Proof The proof is immediate from the above theorem and Theorem 1.2.1.

By the so-called homomorphism theorem, we have the following Fleck's result [18].

Corollary 1.7.2 *Let $\boldsymbol{A} = (S, X, \delta)$ be a strongly connected automaton and H be a normal subgroup of $G(\boldsymbol{A})$. Then $G(\boldsymbol{A})/H$ is isomorphic to a subgroup of $G(\boldsymbol{A}/H)$.*

Corollary 1.7.3 *Let $\boldsymbol{A} = (\widehat{G_1}, X, \delta_\Psi)$ be a regular $(1, G)$-automaton and ξ be a homomorphism of G onto some group. Then $\xi(G)$ is isomorphic to $G(\boldsymbol{A}/\operatorname{Ker}(\xi))$.*

Proof By Theorem 1.7.1, we have $\boldsymbol{A}_\xi \approx \boldsymbol{A}/\operatorname{Ker}(\xi)$. Therefore, $G(\boldsymbol{A}_\xi) \approx G(\boldsymbol{A}/\operatorname{Ker}(\xi))$ holds. On the other hand, by Proposition 1.7.1, $\boldsymbol{A}_\xi$ becomes strongly connected and thus regular. Consequently, we have $\xi(G) \approx G(\boldsymbol{A}_\xi)$, i.e. $\xi(G) \approx G(\boldsymbol{A}/\operatorname{Ker}(\xi))$.

The following Freck's result [18] is an immediate consequence of the above theorem.

Corollary 1.7.4 *Let $\boldsymbol{A} = (S, X, \delta)$ be a strongly connected automaton such that $|S| = |G(\boldsymbol{A})|$ and let H be a normal subgroup of $G(\boldsymbol{A})$. Then $G(\boldsymbol{A})/H \approx G(\boldsymbol{A}/H)$ holds.*

By Theorem 1.2.4 and Proposition 1.7.1, we have the following theorem.

Theorem 1.7.2 *Let $\boldsymbol{A} = (\widehat{G_2}, X, \delta_\Psi)$ be a strongly connected $(2, G)$-automaton and ξ be a homomorphism of G onto some group. Then $\boldsymbol{A}_\xi$ is not regular if and only if there exist some automorphism η of $\xi(G)$, some element $k' \in \xi(G)$, two mappings λ, γ of G into $2^G \setminus \{\emptyset\}$, 2^G means the set of all subsets of G) and two subsets Λ, Γ, $\Gamma \neq \emptyset$ of G, such that $\eta(k') = k'$, $\eta^2(g') = k'g'k'^{-1}$ for any $g' \in \xi(G)$, $\lambda(g) \subseteq \xi^{-1}\eta\xi(g)$ and $\gamma(h) \subseteq \xi^{-1}\eta\xi(h)\xi^{-1}(k')$ for any $g \in \Lambda$ and any $h \in \Gamma$, and $\Psi(a) = \left\{ \begin{pmatrix} g & 0 \\ 0 & x \end{pmatrix}, \begin{pmatrix} 0 & h \\ y & 0 \end{pmatrix} \;\middle|\; x \in \lambda(g),\, y \in \gamma(h),\, g \in \Lambda,\, h \in \Gamma \right\}$.*

Immediately, we have:

Corollary 1.7.5 *Let $\boldsymbol{A} = (S, X, \delta)$ be a strongly connected automaton such that $|S| = 2\,|G(\boldsymbol{A})|$ and let H be a normal subgroup of $G(\boldsymbol{A})$. Then if $\boldsymbol{A}$ is not a permutation automaton, $G(\boldsymbol{A})/H \approx G(\boldsymbol{A}/H)$ holds.*

Remark 1.7.2 The above result will be extended in the following section.

Finally, we apply our method to the proof of an elementary decomposition theorem of automata by Fleck [18].

Theorem 1.7.3 *Let $\boldsymbol{A} = (S, X, \delta)$ be a strongly connected automaton such that $|S| = |G(\boldsymbol{A})|$. Moreover, assume that $G(\boldsymbol{A})$ is the direct product of groups H and K, i.e. $G(\boldsymbol{A}) = H \times K$. Then $\boldsymbol{A} \approx \boldsymbol{A}/H \times \boldsymbol{A}/K$ holds.*

Proof We can assume that $G(\boldsymbol{A}) = G = H \times K$ and $\boldsymbol{A}$ is a regular $(1, G)$-automaton of the form $\boldsymbol{A} = (\widehat{G_1}, X, \delta_\Psi)$.

Now, put $\xi(g) = h$ and $\eta(g) = k$ for any $g = hk \in G$ where $h \in H$ and $k \in K$. Then ξ and η are well defined as homomorphisms of G onto H and onto K, respectively. Moreover, we have $\mathrm{Ker}(\xi) = K$ and $\mathrm{Ker}(\eta) = H$. Consequently, by Theorem 1.7.1, we have $\boldsymbol{A}/H \approx \boldsymbol{A}_\eta$ and $\boldsymbol{A}/K \approx \boldsymbol{A}_\xi$. Put $\boldsymbol{A}_\xi = (\hat{H}_1, X, \delta_{\xi[\Psi]})$, $\boldsymbol{A}_\eta = (\widehat{K_1}, X, \delta_{\eta[\Psi]})$ and $\Psi(a) = (\psi(a))$ for any $a \in X$. Then we have $\xi\,[\Psi]\,(a) = (\xi(\psi(a)))$ and $\eta\,[\Psi]\,(a) = (\eta(\psi(a)))$. By the assumption that $G = H \times K$, we have $\psi(a) = \eta\,[\Psi]\,(a)\xi\,[\Psi]\,(a)$. Consequently, we have $(\eta\,[\Psi] \otimes \xi\,[\Psi])\,(a) = (\psi(a)) = \Psi(a)$. That is, $\eta\,[\Psi] \otimes \xi\,[\Psi] = \Psi$ holds. Therefore, we have $\boldsymbol{A}_\eta \otimes \boldsymbol{A}_\xi = (\widehat{G_1}, X, \delta_{\eta[\Psi]\otimes\xi[\Psi]}) = (\widehat{G_1}, X, \delta_\Psi) = \boldsymbol{A}$. By Theorem 1.6.1, we have $\boldsymbol{A}_\eta \times \boldsymbol{A}_\xi \approx \boldsymbol{A}_\eta \otimes \boldsymbol{A}_\xi$. On the other hand, $\boldsymbol{A}/H \times \boldsymbol{A}/K \approx \boldsymbol{A}_\eta \otimes \boldsymbol{A}_\xi$ holds.

Thus we have $\boldsymbol{A} \approx \boldsymbol{A}/H \times \boldsymbol{A}/K$.

1.8 Prime order case

In the previous section, we dealt with group-matrix type automata. Especially, in Theorem 1.2.5, we gave a necessary and sufficient condition for a strongly connected (n, G)-automaton to be regular. However, the condition seems not easy to handle. In this section we will deal with clear and simple cases, i.e. the cases that the orders of strongly connected group-matrix type automata are prime. Moreover, we apply our results to problems concerning factor automata and direct products of strongly connected automata.

Theorem 1.8.1 *Let $\boldsymbol{A} = (\widehat{G_n}, X, \delta_\Psi)$ be a strongly connected (n, G)-automaton where n is a prime number. Then $\boldsymbol{A}$ is either regular or isomorphic to some group-matrix type automaton of order* 1.

Proof There exist some positive integers p and q such that $|G(\boldsymbol{A})| = p|G|$ and $n|G| = \left|\widehat{G_n}\right| = q|G(\boldsymbol{A})|$. Consequently, we have $n|G| = pq|G|$, namely, $n = pq$. Since n is a prime number, either $(p, q) = (1, n)$ or $(p, q) = (n, 1)$ holds.

Case $(p, q) = (1, n)$. It follows from $|G(\boldsymbol{A})| = |G|$ that $\boldsymbol{A}$ is regular.

Case $(p, q) = (n, 1)$. From $\left|\widehat{G_n}\right| = |G(\boldsymbol{A})|$ and Theorem 1.3.1, it follows that $\boldsymbol{A}$ is isomorphic to some strongly connected $(1, G(\boldsymbol{A}))$-automaton.

Corollary 1.8.1 *Let $\boldsymbol{A}$ be a strongly connected (n, G)-automaton where n is a prime number. Then if $\boldsymbol{A}$ is not a permutation automaton, $\boldsymbol{A}$ is regular.*

Proof This is immediate from the above theorem and Proposition 1.2.2.

Remark 1.8.1 When n is not a prime number, the above is not true.

Example 1.8.1 The following strongly connected $(4, G)$-automaton $\boldsymbol{A}$ is not regular though it is not a permutation automaton:

$\boldsymbol{A} = (\widehat{G_4}, X, \delta_\Psi)$ where $G = \{e, g\}, g^2 = e$ and e is the identity of G and $X = \{a, b, c\}$.

$$\Psi(a) = \begin{pmatrix} g & 0 & 0 & 0 \\ g & 0 & 0 & 0 \\ 0 & 0 & g & 0 \\ 0 & 0 & g & 0 \end{pmatrix}, \quad \Psi(b) = \begin{pmatrix} 0 & e & 0 & 0 \\ e & 0 & 0 & 0 \\ 0 & 0 & 0 & e \\ 0 & 0 & e & 0 \end{pmatrix},$$

$$\Psi(c) = \begin{pmatrix} 0 & 0 & 0 & e \\ 0 & 0 & e & 0 \\ 0 & e & 0 & 0 \\ e & 0 & 0 & 0 \end{pmatrix}.$$

Proof Here, we put the permutation ρ on $\widehat{G_4}$ as follows: $\rho((h,0,0,0)) = (0,0,h,0), \rho((0,0,h,0)) = (h,0,0,0), \rho((0,h,0,0)) = (0,0,0,h)$ and $\rho((0,0,0,h)) = (0,h,0,0)$ for any $h \in G$.

Then we have $\rho \in G(\boldsymbol{A})$. But, since we cannot represent ρ in the form $\rho = \rho_g$ $(g \in G)$, $G(\boldsymbol{A}) \approx G$ does not hold. Thus $\boldsymbol{A}$ is not regular.

Now, let us determine the automorphism group $G(\boldsymbol{A})$ of $\boldsymbol{A}$ in the case that $\boldsymbol{A}$ is not regular. For this problem, we have the following result.

Theorem 1.8.2 *Let $\boldsymbol{A} = (\widehat{G_n}, X, \delta_\Psi)$ be a strongly connected (n, G)-automaton where n is a prime number. Then $G(\boldsymbol{A})$ is isomorphic to $\Psi(X^*)$ if $\boldsymbol{A}$ is not regular.*

Proof By Theorem 1.8.1, there exists a strongly connected $(1, G(\boldsymbol{A}))$-automaton $\boldsymbol{A}'$ isomorphic to $\boldsymbol{A}$. Therefore, we have $G(\boldsymbol{A}) \approx G(\boldsymbol{A}')$ and $C(\boldsymbol{A}) \approx C(\boldsymbol{A}')$. On the other hand, $C(\boldsymbol{A}') \approx G(\boldsymbol{A}')$ and $C(\boldsymbol{A}) \approx \Psi(X^*)$ hold. Consequently, we have $G(\boldsymbol{A}) \approx \Psi(X^*)$.

1.9 Regularities

We will consider, in this section, the regularities of strongly connected (n, G)-automata. First, we give the following lemma.

Lemma 1.9.1 *Let $\boldsymbol{A} = (S, X, \delta)$ be a strongly connected automaton isomorphic to some $(1, G)$-automaton. Furthermore, assume that there exist some $s_0 \in S$ and $x, y \in X^*$ such that $\delta(s_0, x) = \delta(s_0, y)$. Then for any $s \in S$ we have $\delta(s, x) = \delta(s, y)$.*

Proof Notice that we can assume that $\boldsymbol{A}$ itself is a strongly connected $(1, G)$-automaton of the form $\boldsymbol{A} = (\widehat{G_1}, X, \delta_\Psi)$, i.e. $S = \widehat{G_1}$ and $\delta = \delta_\Psi$. Then $\delta(s_0, x) = \delta(s_0, y)$ implies $s_0\Psi(x) = s_0\Psi(y)$, and thus we have $\Psi(x) = \Psi(y)$. Therefore, for any $s \in S$ $(= \widehat{G_1} = G)$ we have $s\Psi(x) = s\Psi(y)$.

The following is the main theorem in this section.

Theorem 1.9.1 *Let $\boldsymbol{A} = (\widehat{G_n}, X, \delta_\Psi)$ be a strongly connected (n, G)-automaton where n is a prime number. Then $\boldsymbol{A}$ is not regular if and only if the following two conditions are satisfied:*

(1) For any $Y \in \Psi(X^*)$ there exist some $\sigma \in S(n)$ and $g_{ij} \in G, i, j = 1, 2, \ldots, n$ such that $Y = \left(g_{pq}e_{p\sigma(q)}\right)$.

(2) Let $Y = (y_{pq})$ and $Z = (z_{pq})$ be two elements in $\Psi(X^*)$. Then if there exist some $i, j = 1, 2, \ldots, n$ such that $y_{ij} = z_{ij} \neq 0$, we have $Y = Z$.

Proof ($\Rightarrow$) The proof of (1) is immediate from Theorem 1.8.1 and Proposition 1.5.3.

Now, we prove (2). Let y and z be elements in X^* such that $\Psi(y) = Y$ and $\Psi(z) = Z$. Furthermore, we put $\widehat{e_i} = (e_{pi})$. Then we have $\delta_\Psi(\widehat{e_i}, y) = (y_{ij}e_{pj}) = (z_{ij}e_{pj}) = \delta_\Psi(\widehat{e_i}, z)$. By Lemma 1.9.1, $\delta_\Psi(\widehat{g}, y) = \delta_\Psi(\widehat{g}, z)$ holds for any $\widehat{g} \in \widehat{G_n}$. Consequently, we have $Y = Z$.

($\Leftarrow$) Notice that $H = \Psi(X^*)$ forms a group from (1). Here, we put $\boldsymbol{A}' = (\widehat{H_1}, X, \delta_{\Psi'})$ where $\Psi'(a) = \Psi(a)$ for any $a \in X$.

Now, let us prove that $\boldsymbol{A}$ is isomorphic to $\boldsymbol{A}'$.

Let ρ be a mapping of $\widehat{G_n}$ into H such that

$$\rho((0, \ldots, 0, \overset{i\text{-th}}{g}, 0, \ldots, 0)) = \begin{pmatrix} 0 & 0 & \cdots & \overset{i\text{-th}}{g} & \cdots & 0 \\ * & * & *** & * & *** & * \\ * & * & *** & * & *** & * \end{pmatrix} \in H$$

for any $g \in G$ and any $i = 1, 2, \ldots, n$.

By (2) and by the strong connectedness of $\boldsymbol{A}$, we can verify that ρ is uniquely determined as a surjective mapping of $\widehat{G_n}$ onto $\widehat{H_1}$ $(= H)$.

We will prove that $\rho(\delta_\Psi(\widehat{g}, a)) = \delta_{\Psi'}(\rho(\widehat{g}), a)$ for any $\widehat{g} \in \widehat{G_n}$ and any $a \in X$. Put $\widehat{g} = (ge_{pi}), g \in G$ and $\Psi(a) = (y_{pq})$. Then we have

$\widehat{g}\Psi(a) = \left(\sum_{k=1}^{n} g e_{ki} y_{kp}\right) = (g y_{ip})$. Here, assume that $y_{ip'} \neq 0, 1 \leq p' \leq n$. Thus we have $\widehat{g}\Psi(a) = (g y_{ip'} e_{pp'})$. Cosequently, we have:

$$\rho(\delta_\Psi(\widehat{g}, a)) = \begin{pmatrix} 0 & 0 & \cdots & 0 & \overset{p'\text{-th}}{g y_{ip'}} & 0 & \cdots & 0 \\ * & * & *** & * & * & * & *** & * \\ * & * & *** & * & * & * & *** & * \end{pmatrix}.$$

On the other hand, we have $\delta_{\Psi'}(\rho(\widehat{g}), a) = \rho(\widehat{g})\Psi'(a) = \rho(\widehat{g})\Psi(a)$

$$= \begin{pmatrix} 0 & \cdots & 0 & \overset{i\text{-th}}{g} & 0 & \cdots & 0 \\ * & *** & * & * & * & *** & * \\ * & *** & * & * & * & *** & * \end{pmatrix} (y_{pq})$$

$$= \begin{pmatrix} 0 & \cdots & 0 & \overset{p'\text{-th}}{g y_{ip'}} & 0 & \cdots & 0 \\ * & *** & * & * & * & *** & * \\ * & *** & * & * & * & *** & * \end{pmatrix}.$$

By (2), $\rho(\delta_\Psi(\widehat{g}, a)) = \delta_{\Psi'}(\rho(\widehat{g}), a)$ holds. Thus we have $\boldsymbol{A} \approx \boldsymbol{A}'$. Consequently, $\boldsymbol{A}'$ is a strongly connected $(1, H)$-automaton. Therefore, $G(\boldsymbol{A}') \approx H$ holds. On the other hand, since $G(\boldsymbol{A}) \approx G(\boldsymbol{A}')$, we have $|G(\boldsymbol{A})| = |H|$. By (2) and by the strong connectedness of $\boldsymbol{A}$, we can obtain easily $|H| = n\,|G|$. Thus we have $|G(\boldsymbol{A})| = n\,|G|$. Since $n > 1$, we have $|G(\boldsymbol{A})| > |G|$.

Consequently, we have $G(\boldsymbol{A}) \not\approx G$, and this means that $\boldsymbol{A}$ is not regular.

Example 1.9.1 The following $(3, G)$-automaton $\boldsymbol{A}$ satisfies the conditions of Theorem 1.9.1: $\boldsymbol{A} = (\widehat{G_3}, X, \delta_\Psi)$ where $G = \{e, g\}, g^2 = e$ and e is the identity of G, and $X = \{a, b\}$.

$$\Psi(a) = \begin{pmatrix} g & 0 & 0 \\ 0 & g & 0 \\ 0 & 0 & g \end{pmatrix} \text{ and } \Psi(b) = \begin{pmatrix} 0 & 0 & e \\ e & 0 & 0 \\ 0 & e & 0 \end{pmatrix}.$$

Thus $\boldsymbol{A}$ is not regular, and we have

$$G(\boldsymbol{A}) \approx \left\{ \begin{pmatrix} 0 & 0 & e \\ e & 0 & 0 \\ 0 & e & 0 \end{pmatrix}, \begin{pmatrix} 0 & 0 & a \\ a & 0 & 0 \\ 0 & a & 0 \end{pmatrix}, \begin{pmatrix} 0 & e & 0 \\ 0 & 0 & e \\ e & 0 & 0 \end{pmatrix}, \begin{pmatrix} 0 & g & 0 \\ 0 & 0 & g \\ g & 0 & 0 \end{pmatrix}, \begin{pmatrix} e & 0 & 0 \\ 0 & e & 0 \\ 0 & 0 & e \end{pmatrix}, \begin{pmatrix} g & 0 & 0 \\ 0 & g & 0 \\ 0 & 0 & g \end{pmatrix} \right\}.$$

Now, put $s_1 = (e, 0, 0)$, $s_2 = (g, 0, 0)$, $s_3 = (0, e, 0)$, $s_4 = (0, g, 0)$, $s_5 = (0, 0, e)$ and $s_6 = (0, 0, g)$. Then we obtain the following state transition diagram of $\boldsymbol{A}$:

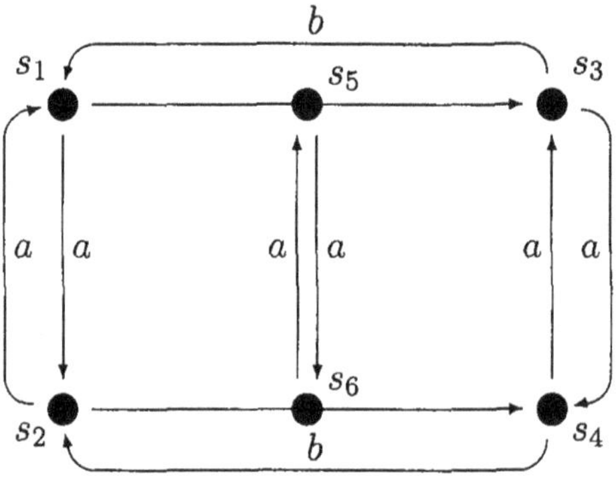

Figure 1.5: State transition diagram of $\boldsymbol{A}$

1.10 Application to factor automata

In this short section, we apply our results to the problem developed in Section 1.7.

Let $\boldsymbol{A} = (\widehat{G_n}, X, \delta_\Psi)$ be a strongly connected (n, G)-automaton where n is a prime number. Furthermore, assume that ξ is a homomorphism of G onto some group. Then as mentioned in Section 1.7, $\boldsymbol{A}_\xi = (\widehat{\xi(G)}_n, X, \delta_{\xi[\Psi]})$ is a strongly connected $(n, \xi(G))$-automaton. Moreover, by means of Theorem 1.9.1, we can decide whether $\boldsymbol{A}_\xi$ is regular or not.

For example, we have the following theorem.

Theorem 1.10.1 *Let* $\boldsymbol{A}$ *be a strongly connected* (n, G)*-automaton where* n *is a prime number. Furthermore, assume that* ξ *is a homomorphism of* G *onto some group. Then if* $\boldsymbol{A}$ *is not a permutation automaton,* $\boldsymbol{A}_\xi$ *is regular.*

Corollary 1.10.1 *Let* $\boldsymbol{A} = (S, X, \delta)$ *be a strongly connected automaton such that* $|S| = n\,|G(\boldsymbol{A})|$ *where* n *is a prime number. Furthermore, assume that* H *is a normal subgroup of* $G(\boldsymbol{A})$*. Then if* $\boldsymbol{A}$ *is not a permutation automaton, we have* $G(\boldsymbol{A})/H \approx G(\boldsymbol{A}/H)$*.*

1.11 Application to direct products

In this section, we provide a relationship between the direct product of automata and its automorphism group.

Theorem 1.11.1 *Let* $\boldsymbol{A} = (S, X, \delta)$ *and* $\boldsymbol{B} = (T, X, \gamma)$ *be strongly connected automata such that* $|S| = |G(\boldsymbol{A})|$ *and* $|T| = n\,|G(\boldsymbol{B})|$*, where* n *is a prime number. Furthermore, assume that* $\boldsymbol{A} \times \boldsymbol{B}$ *is strongly connected. Then if one of the following two conditions is satisfied,* $G(\boldsymbol{A} \times \boldsymbol{B}) \approx G(\boldsymbol{A}) \times G(\boldsymbol{B})$ *holds.*

(1) $\boldsymbol{B}$ *is not a permutation automaton.*

(2) $|G(\boldsymbol{A})|$ *and* $(n-1)!$ *are relatively prime, and so are* $|G(\boldsymbol{A})|$ *and* $|G(\boldsymbol{B})|$*.*

Proof By Theorem 1.3.1, there exist a regular $(1, G(\boldsymbol{A}))$-automaton $\boldsymbol{A}' = (\widehat{G(\boldsymbol{A})}_1, X, \delta_\Psi)$ and also a regular $(n, G(\boldsymbol{B}))$-automaton $\boldsymbol{B}' = (\widehat{G(\boldsymbol{B})}_n, X, \delta_\Pi)$ such that $\boldsymbol{A} \approx \boldsymbol{A}'$ and $\boldsymbol{B} \approx \boldsymbol{B}'$. Hence, by Theorem 3.6.1, $\boldsymbol{A}' \times \boldsymbol{B}' \approx \boldsymbol{A}' \otimes \boldsymbol{B}'$ holds. Here, $\boldsymbol{A}' \otimes \boldsymbol{B}' = ((\widehat{G(\boldsymbol{A}) \times G(\boldsymbol{B})})_n, X, \delta_{\Psi \otimes \Pi})$. Since $\boldsymbol{A} \times \boldsymbol{B} \approx \boldsymbol{A}' \otimes \boldsymbol{B}'$, we have $G(\boldsymbol{A} \times \boldsymbol{B}) \approx G(\boldsymbol{A}' \otimes \boldsymbol{B}')$. Therefore, to prove the theorem, it is enough to verify that $\boldsymbol{A}' \otimes \boldsymbol{B}'$ is regular. First, notice that $\boldsymbol{A}' \otimes \boldsymbol{B}'$ is strongly connected as well as $\boldsymbol{A} \times \boldsymbol{B}$.

Proof of the regularity of $\boldsymbol{A}' \otimes \boldsymbol{B}'$ Case 1. Consider the case that $\boldsymbol{B}'$ is not a permutation automaton. Then $\boldsymbol{A}' \otimes \boldsymbol{B}'$ is not also a permutation automaton. Therefore, in this case, by Corollary 1.8.1, $\boldsymbol{A}' \otimes \boldsymbol{B}'$ becomes regular.

Case 2. Now, assume that $\boldsymbol{B}'$ is a permutation automaton. Then obviously, so is $\boldsymbol{A}' \otimes \boldsymbol{B}'$. First, prove that there exist an element $x \in X^*$ and two positive integers $i, j = 1, 2, \ldots, n$, $i \neq j$ such that $\Pi(x) = (y_{pq}), y_{ii} = e$ and $y_{jj} \neq e$ where e is the identity of $G(\boldsymbol{B})$.

Proof of the above assertion Since $\boldsymbol{B}'$ is regular, by Theorem 1.9.1, there exist some elements y, z in X^* and positive integers $i, i', j, j' = 1, 2, \ldots, n$, $i \neq j$ such that $\Pi(y) = (y_{pq})$, $\Pi(z) = (z_{pq})$, $y_{ii'} = z_{ii'} \neq 0$ and $y_{jj'} \neq z_{jj'}$. Since $\boldsymbol{B}'$ is a permutation automaton, there exists some positive integer $m > 1$ such that $\Pi(y)^m = \Pi(y^m) = (e_{pq})$. Here, we put $x = zy^{m-1}$. Thus $\Pi(x) = (y_{pq})$, $y_{ii} = e$ and $y_{jj} \neq e$.

We have the following two cases.

(i) $y_{jj} = 0$.

(ii) $e \neq y_{jj} \neq 0$.

Case (i). Put $\Psi(x) = g \in G(\boldsymbol{A})$ $(= \widetilde{G(\boldsymbol{A})}_1)$. Then $(\Psi \otimes \Pi)(x) = (gy_{pq})$. Since $\boldsymbol{A}' \otimes \boldsymbol{B}'$ is a permutation automaton, there exist some $f_{pq} \in G(\boldsymbol{B}), p, q = 1, 2, \ldots, n$ and a permutation τ on $\{1, 2, 3, \ldots, n\}$ such that $\tau(i) = i$, $\tau(j) \neq j$ and $(\Psi \otimes \Pi)(x) = (gy_{pq}) = (gf_{pq}e_{p\tau(q)})$. Now, we calculate $(u_{pq}) = (\Psi \otimes \Pi)(x^t)$ where $t = |G(\boldsymbol{A})|$. From the assumption that $|G(\boldsymbol{A})|$ and $(n-1)!$ are relatively prime, $u_{ii} = e$ and $u_{jj} = 0$. On the other hand, $(\Psi \otimes \Pi)(\varepsilon) = (e_{pq})$. Therefore, by Theorem 1.9.1, $\boldsymbol{A}' \otimes \boldsymbol{B}'$ becomes regular.

Case (ii). Put $\Psi(x) = g \in G(\boldsymbol{A})$ $(= \widetilde{G(\boldsymbol{A})}_1)$. Then $(\Psi \otimes \Pi)(x) = (gy_{pq})$ holds. Let t be a positive integer and put $(\Psi \otimes \Pi)(x^t) = (v_{pq})$. Then we have $v_{ii} = g^t$ and $v_{jj} = g^t y_{jj}^t$. Now, we put $t = |G(\boldsymbol{A})|$. Since $g \in G(\boldsymbol{A})$, $e \neq y_{jj} \in G(\boldsymbol{B})$ and, $|G(\boldsymbol{A})|$ and $|G(\boldsymbol{B})|$ are relatively prime, we have $v_{ii} = e$ and $v_{jj} \neq e$. On the other hand, $(\Psi \otimes \Pi)(\varepsilon) = (e_{pq})$. Therefore, by Theorem 1.9.1, $\boldsymbol{A}' \otimes \boldsymbol{B}'$ becomes regular.

Remark 1.11.1 In the above theorem, neither of two conditions of (2) can be eliminated. First, consider the case that $|G(\boldsymbol{A})|$ and $(n-1)!$ are not relatively prime.

Example 1.11.1 Let $H = \{e, h\}, h^2 = e$, $K = \{e\}$ and $H \times K = \{e, h\}$ where e is the identity of $H \times K$. Put $X = \{a, b\}$, $\Psi(a) = h$, $\Psi(b) = e$, $\Pi(a) = \begin{pmatrix} e & 0 & 0 \\ 0 & 0 & e \\ 0 & e & 0 \end{pmatrix}$ and $\Pi(b) = \begin{pmatrix} 0 & e & 0 \\ 0 & 0 & e \\ e & 0 & 0 \end{pmatrix}$.

Then $\boldsymbol{A} = (\widehat{H_1}, X, \delta_\Psi)$ and $\boldsymbol{B} = (\widehat{K_3}, X, \delta_\Pi)$ are regular group-matrix type automata. Notice that $|G(\boldsymbol{A})| = |H| = 2$ and $(n-1)! = 2$. Moreover, $\boldsymbol{A} \otimes \boldsymbol{B} = ((\widehat{H \times K})_3, X, \delta_{\Psi\otimes\Pi})$.

Here, $(\Psi \otimes \Pi)(a) = \begin{pmatrix} h & 0 & 0 \\ 0 & 0 & h \\ 0 & h & 0 \end{pmatrix} \in (\widetilde{H \times K})_3$ and $(\Psi \otimes \Pi)(b) = \begin{pmatrix} 0 & e & 0 \\ 0 & 0 & e \\ e & 0 & 0 \end{pmatrix} \in (\widetilde{H \times K})_3$.

Then we can calculate easily:

$$\Psi \otimes \Pi)(X^*) = \left\{ \begin{pmatrix} e & 0 & 0 \\ 0 & e & 0 \\ 0 & 0 & e \end{pmatrix}, \begin{pmatrix} 0 & e & 0 \\ 0 & 0 & e \\ e & 0 & 0 \end{pmatrix}, \begin{pmatrix} 0 & 0 & e \\ e & 0 & 0 \\ 0 & e & 0 \end{pmatrix}, \right.$$
$$\left. \begin{pmatrix} 0 & 0 & h \\ 0 & h & 0 \\ h & 0 & 0 \end{pmatrix}, \begin{pmatrix} 0 & h & 0 \\ h & 0 & 0 \\ 0 & 0 & h \end{pmatrix}, \begin{pmatrix} h & 0 & 0 \\ 0 & 0 & h \\ 0 & h & 0 \end{pmatrix} \right\}.$$

From the above, we can see that $\boldsymbol{A} \otimes \boldsymbol{B}$ is strongly connected, but not regular. Thus $G(\boldsymbol{A} \times \boldsymbol{B}) \approx G(\boldsymbol{A}) \times G(\boldsymbol{B})$ does not hold.

Now, we consider the case that $|G(\boldsymbol{A})|$ and $|G(\boldsymbol{B})|$ are not relatively prime.

Example 1.11.2 For the following automata $\boldsymbol{A}$ and $\boldsymbol{B}$, $G(\boldsymbol{A} \times \boldsymbol{B}) \approx G(\boldsymbol{A}) \times G(\boldsymbol{B})$ does not hold.

$H = \{e, g\}, g^2 = e, K = \{e, h\}, h^2 = e, gh = hg$ and e is the identity of $H \times K$.

$H \times K = \{e, g, h, gh\}, \boldsymbol{A} = (\widehat{H_1}, X, \delta_\Psi)$ and $\boldsymbol{B} = (\widehat{K_2}, X, \delta_\Pi)$ where $X = \{a, b\}$.

$$\Psi(a) = e \in H \ \ (= \widetilde{H_1}),\ \Psi(b) = g \in H \ \ (= \widetilde{H_1}).$$

$$\Pi(a) = \begin{pmatrix} 0 & h \\ e & 0 \end{pmatrix} \in \widetilde{K_2},\ \Pi(b) = \begin{pmatrix} 0 & e \\ e & 0 \end{pmatrix} \in \widetilde{K_2}.$$

Then we have $G(\boldsymbol{A}) \approx H$ and $G(\boldsymbol{B}) \approx K$. Moreover, $|G(\boldsymbol{A})| = |G(\boldsymbol{B})| = 2$. On the other hand, $\boldsymbol{A} \otimes \boldsymbol{B} = ((\widehat{H \times K})_2, X, \delta_{\Psi\otimes\Pi})$ where

$$\Psi \otimes \Pi(a) = \begin{pmatrix} 0 & h \\ e & 0 \end{pmatrix} \text{ and } \Psi \otimes \Pi(b) = \begin{pmatrix} 0 & g \\ g & 0 \end{pmatrix}.$$

It is easy to verify that $\boldsymbol{A} \otimes \boldsymbol{B}$ is not regular. Thus $G(\boldsymbol{A} \times \boldsymbol{B}) \approx G(\boldsymbol{A}) \times G(\boldsymbol{B})$ does not hold.

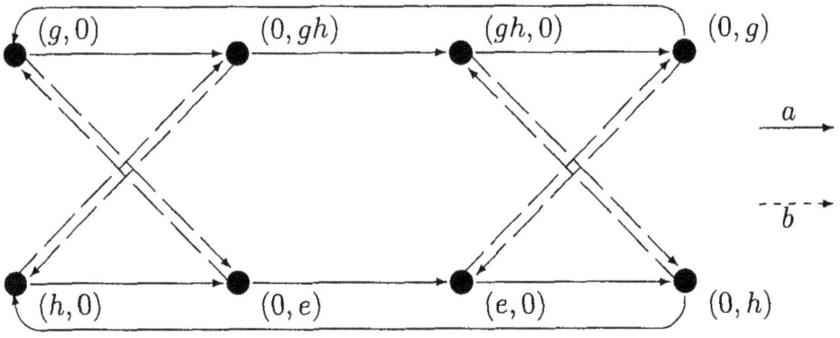

Figure 1.6: State transition diagram of $\boldsymbol{A} \otimes \boldsymbol{B}$

Chapter 2

General Automata

In the previous chapter, we discussed the automorphism groups of strongly connected automata. To this end, we introduced the representations of automata, called *group-matrix type automata*. However, the strong connectedness of automata was indispensable for these sorts of representations. Thus we felt the necessity of introducing some other representations of automata connecting with the automorphism groups of general automata. In the following sections, we will attempt to generalize our results developed in the previous chapter for the case of general automata. For this purpose, we will introduce *generalized group-matrix type automata* as representations of general automata and we will also provide several properties of these sorts of automata.

The contents of this chapter are based on the results in [33].

2.1 Generalized representations

In this section, we will define a *generalized group-matrix type automaton*. To this end, we will introduce the notion of the *SP-condition* for a finite group G and a family of subgroups of G.

Lemma 2.1.1 *Let H and K be subgroups of a given finite group G, respectively. Moreover, let g be an element in G. Then $H \subseteq gKg^{-1}$ holds if and only if $(g'H)(gK) \subseteq g'gK$ holds for any $g' \in G$.*

Proof (⇒) $(g'H)(gK) \subseteq g'gKg^{-1}gK = g'gK$.

(⇐) Let $g' = e$ where e is the identity of G. Then $HgK \subseteq gK$. Hence $Hg \subseteq gKg^{-1}$.

Definition 2.1.1 Let G be a finite group and let $H^{(n)} = \{H_1, H_2, H_3, \ldots, H_n\}$ be a family of subgroups of G satisfying the following condition:

$$\bigcap_{g\in G}(\bigcap_{i=1}^{n} gH_i g^{-1}) = \{e\}.$$

Henceforth, we call this condition the *separation condition* (or in short, the *SP-condition*).

Now, we define generalized group-matrices.

A matrix $(g_{pq}H_q)$ is called a *generalized group-matrix of order n over* $(H^{(n)}, G)$ and denoted $(g_{pq}H_q) \in \widetilde{G}(H^{(n)})$ if the following two conditions are satisfied:

(1) $(g_{pq}) \in \widetilde{G_n}$.

(2) If $g_{ij} \neq 0$, then $H_i \subseteq g_{ij}H_j g_{ij}^{-1}$.

Definition 2.1.2 Let H and K be subgroups of a given finite group G, respectively. We introduce two operations $(\cdot)$ and $(+)$ as follows:

(1) For $g, h \in G$, we put $(gH) \cdot (hK) = ghK$ if it is well-defined.

(2) For any $g, h \in G$, we define $0 \cdot (hK) = (gH) \cdot 0 = 0 \cdot 0 = 0$ where $0 \cdot H = 0 \cdot K = 0$.

(3) For any $g \in G$, we define $(gH) + 0 = 0 + (gH) = gH$ and $0 + 0 = 0$.

(4) For any $g, h \in G$, we do not define $(gH) + (hK)$.

We will use sometimes the notations $(gH)(hK)$ and $\sum_{i=1}^{s} g_i H_i$ instead of $(gH) \cdot (hK)$ and $g_1H_1 + g_2H_2 + \cdots + g_sH_s$. Notice that the sum $\sum_{i=1}^{s} g_i H_i$ is defined only if at most one of $g_iH_i, i = 1, 2, \ldots, s$ is nonzero.

Proposition 2.1.1 *Let H and K be subgroups of a given finite group G, respectively. Moreover, assume that there exists an element $h \in K$ such that $H \subseteq hKh^{-1}$. Then for any $g \in G$ we can define $(gH)(hK)$.*

Proof Obvious from Lemma 2.1.1.

Definition 2.1.3 Let $(g_{pq}H_q)$ and $(h_{pq}H_q)$ be elements in $\widetilde{G}(H^{(n)})$. Then we define the multiplication $(g_{pq}H_q)(h_{pq}H_q)$ as follows:

$$(g_{pq}H_q)(h_{pq}H_q) = \left(\sum_{k=1}^{n}(g_{pk}H_k)(h_{kq}H_q)\right) = \left(\sum_{k=1}^{n} g_{pk}h_{kq}H_q\right).$$

In this case, $(g_{pq}H_q)(h_{pq}H_q)$ is well-defined and it is contained in $\widetilde{G}(H^{(n)})$.

Adequacy of the definition It is enough to show that, for $g_{ik}, h_{kj} \in G$, $(g_{ik}H_k)(h_{kj}H_j)$ is well-defined, i.e. we have to show that $H_k \subseteq h_{kj}H_jh_{kj}^{-1}$.

However, this is obvious from the fact that $(h_{pq}H_q) \in \widetilde{G}(H^{(n)})$.

Proof of $(g_{pq}H_q)(h_{pq}H_q) \in \widetilde{G}(H^{(n)})$ Let $(f_{pq}H_q) = (g_{pq}H_q)(h_{pq}H_q)$. Then $f_{ij} = \sum_{k=1}^{n} g_{ik}h_{kj}$ holds for any $i, j = 1, 2, \ldots, n$.

It follows that $(f_{pq}) \in \widetilde{G_n}$ from the fact that $(g_{pq}H_q), (h_{pq}H_q) \in \widetilde{G}(H^{(n)})$. Therefore, it is enough to prove that, if $f_{ij} \neq 0$, then $H_i \subseteq f_{ij}H_jf_{ij}^{-1}$.

First, notice that, for $f_{ij} \neq 0$, we can find some $r = 1, 2, \ldots, n$ such that $f_{ij} = \sum_{k=1}^{n} g_{ik}h_{kj} = g_{ir}h_{rj}$. From the fact that $g_{ir} \neq 0$ and $h_{rj} \neq 0$, we have $H_i \subseteq g_{ir}H_rg_{ir}^{-1}$ and $H_r \subseteq h_{rj}H_jh_{rj}^{-1}$. Thus we have $H_i \subseteq (g_{ir}h_{rj})H_j(g_{ir}h_{rj})^{-1}$,i.e. $H_i \subseteq f_{ij}H_jf_{ij}^{-1}$.

This completes the proof of $(g_{pq}H_q)(h_{pq}H_q) \in \widetilde{G}(H^{(n)})$.

Definition 2.1.4 Let $H^{(n)} = \{H_1, H_2, H_3, \ldots, H_n\}$ be a family of subgroups of a given finite group G satisfying the SP-condition. Then a vector (f_pH_p) is called a *generalized group-vector of order n over $(H^{(n)}, G)$* and denoted $(f_pH_p) \in \widehat{G}(H^{(n)})$ if $(f_p) \in \widehat{G_n}$ holds.

For any $(f_pH_p) \in \widehat{G}(H^{(n)})$ and any $(g_{pq}H_q) \in \widetilde{G}(H^{(n)})$, we define the following multiplication:

$$(f_pH_p)(g_{pq}H_q) = \left(\sum_{k=1}^{n}(f_kH_k)(g_{kp}H_p)\right) = \left(\sum_{k=1}^{n} f_k g_{kp} H_p\right).$$

Under this operation, we obtain $(f_pH_p)(g_{pq}H_q) \in \widehat{G}(H^{(n)})$.

Definition 2.1.5 Let $H^{(n)} = \{H_1, H_2, H_3, \ldots, H_n\}$ be a family of subgroups of a given finite group G satisfying the SP-condition. An automaton $\boldsymbol{A} = (\widehat{G}(H^{(n)}), X, \delta_\Psi)$ is called a *generalized group-matrix type automaton of order n over* $(H^{(n)}, G)$ (or in short, an $(H^{(n)}, G)$-*automaton*) if $\boldsymbol{A}$ consists of the following data:

(1) $\widehat{G}(H^{(n)})$ is the set of states.

(2) X is a set of inputs.

(3) δ_Ψ is a state transition function and it is defined by
$\delta_\Psi((f_pH_p), a) = (f_pH_p)\Psi(a)$ for $(f_pH_p) \in \widehat{G}(H^{(n)})$ and $a \in X$ where Ψ is a mapping of X into $\widetilde{G}(H^{(n)})$.

Remark 2.1.1 The mapping Ψ can be extended to the mapping of X^* into $\widetilde{G}(H^{(n)})$ as follows:

$$\Psi(\epsilon) = (e_{pq}H_q) \text{ and } \Psi(xy) = \Psi(x)\Psi(y) \text{ for any } x, y \in X^*.$$

In this case, we can easily see that $\delta_\Psi((f_pH_p), x) = (f_pH_p)\Psi(x)$ holds for any $(f_pH_p) \in \widehat{G}(H^{(n)})$ and any $x \in X^*$.

2.2 Some properties

In this section, some properties of generalized group-matrix type automata are described. First, notice that, in the definition of generalized group-matrix type automaton, if $H_i = \{e\}$ for any $i = 1, 2, \ldots, n$ where $H^{(n)} = \{H_1, H_2, H_3, \ldots, H_n\}$, then an $(H^{(n)}, G)$-automaton is regarded as an (n, G)-automaton.

Proposition 2.2.1 *Let* $\boldsymbol{A}$ *be an* $(H^{(1)}, G)$*-automaton where* $H^{(1)} = \{H\}$*. Then if* H *is a normal subgroup of* G*,* $\boldsymbol{A}$ *is a* $(1, G)$*-automaton.*

Proof By the SP-condition, we have $\bigcap_{g \in G} gHg^{-1} = \{e\}$. Since H is a normal subgroup of G, $gHg^{-1} = H$ holds for any $g \in G$. Thus we have $H = \{e\}$. This means that $\boldsymbol{A}$ is a $(1, G)$-automaton.

Remark 2.2.1 In the above, the assertion of the proposition is not true when H is not a normal subgroup of G.

For example, let G be the alternating group $A(5)$ of degree 5 and let H be a subgroup of G such that $\{e\} \neq H \neq G$. Then since $A(5)$ is a simple group, the normal subgroup $\bigcap_{g \in G} gHg^{-1}$ must be equal to $\{e\}$. Thus $H^{(1)}$ satisfies the SP-condition. Now, put $X = \{a\}$ and $\Psi(a) = (H)$. Then it is easily seen that $\boldsymbol{A} = (\widehat{G}(H^{(1)}), X, \delta_{\Psi})$ is an $(H^{(1)}, G)$-automaton. Therefore, there exists an $(H^{(1)}, G)$-automaton which is not a $(1, G)$-automaton.

Proposition 2.2.2 *A strongly connected* $(H^{(n)}, G)$*-automaton is an* (n, G)*-automaton.*

Proof Let $\boldsymbol{A}$ be a strongly connected $(H^{(n)}, G)$-automaton with $H^{(n)} = \{H_1, H_2, H_3, \ldots, H_n\}$. By the strong connectedness of $\boldsymbol{A}$, for any $g, h \in G$ and any $i, j = 1, 2, \ldots, n$, there exists some $Y \in \Psi(X^*)$ such that $(0, \ldots, 0, gH_i, 0, \ldots, 0)Y = (0, \ldots, 0, hH_j, 0, \ldots, 0)$. From the above, we can see that, for any $f \in G$ and any $i, j = 1, 2, \ldots, n$, $H_i \subseteq fH_jf^{-1}$ holds. Thus we have $H_i \subseteq \bigcap_{f \in G} (\bigcap_{j=1}^{n} fH_jf^{-1})$ for any $i = 1, 2, \ldots, n$. By the SP-condition, we have $H_i = \{e\}$ for any $i = 1, 2, \ldots, n$.

Consequently, $\boldsymbol{A}$ is an (n, G)-automaton.

Remark 2.2.2 Similar results can be obtained for a weakly connected automaton and a singly generated automaton. For these sorts of automata, see Bavel [6].

Theorem 2.2.1 *Let G be a finite group and let $\boldsymbol{A} = (\widehat{G}(H^{(n)}), X, \delta_\Psi)$ be an $(H^{(n)}, G)$-automaton. Then G is isomorphic to a subgroup of $G(\boldsymbol{A})$.*

Proof For any $g \in G$, we define the following mapping ρ_g of $\widehat{G}(H^{(n)})$ onto itself:

$$\rho_g((f_pH_p)) = (gf_pH_p) \text{ for any } (f_pH_p) \in \widehat{G}(H^{(n)}).$$

First, it is easily seen that ρ_g is well-defined as a permutation on $\widehat{G}(H^{(n)})$.

Now we show that $\rho_g(\delta_\Psi((0,\dots,0,g_iH_i,0,\dots,0),a)) = \delta_\Psi(\rho_g((0,\dots,0,g_iH_i,0,\dots,0)),a)$ holds for any $a \in X$ and any $g_i \in G, i = 1,2,\dots,n$.

Let a be an element in X and let $\Psi(a) = (h_{pq}H_q)$. Assume that $h_{ij} \neq 0$. Notice that $j \geq 1$ is determined uniquely with respect to i. Then we have $\delta_\Psi((0,\dots,0,g_iH_i,0,\dots,0),a) = (0,\dots,0,g_iH_i,0,\dots,0)\Psi(a) = (0,\dots,0,g_ih_{ij}H_j,0,\dots,0)$. Thus we have $\rho_g(\delta_\Psi((0,\dots,0,g_iH_i,0,\dots,0),a)) = (0,\dots,0,gg_ih_{ij}H_j,0,\dots,0)$.

On the other hand, we have $\delta_\Psi(\rho_g((0,\dots,0,g_iH_i,0,\dots,0),a) = \delta_\Psi((0,\dots,0,gg_iH_i,0,\dots,0),a) = (0,\dots,0,gg_iH_i,0,\dots,0)\Psi(a) = (0,\dots,0,gg_ih_{ij}H_j,0,\dots,0)$.

Hence we have $\rho_g(\delta_\Psi((0,\dots,0,g_iH_i,0,\dots,0),a)) = \delta_\Psi(\rho_g((0,\dots,0,g_iH_i,0,\dots,0)),a)$. Since ρ_g is a permutation on $\widehat{G}(H^{(n)})$, we have $\rho_g \in G(\boldsymbol{A})$.

Next, let us prove that the mapping $g \to \rho_g$ is an isomorphism of G into $G(\boldsymbol{A})$. Since it is almost obvious that the mapping $g \to \rho_g$ is a homomorphism, we prove only that $g \to \rho_g$ is an injective mapping.

Assume that $\rho_g = \rho_h$ holds for $g,h \in G$. Then for any $i = 1,2,\dots,n$ and any $f \in G$, we have $\rho_g((0,\dots,0,fH_i,0,\dots,0)) = \rho_h((0,\dots,0,fH_i,0,\dots,0))$. Thus $gfH_i = hfH_i$ holds for any $i = 1,2,\dots,n$ and any $f \in G$, and this implies $h^{-1}g \in \bigcap_{f\in G}(\bigcap_{i=1}^{n} fH_if^{-1})$. By the SP-condition, we have $h = g$.

This means that $g \to \rho_g$ is an injective mapping, and hence we complete the proof of the Theorem.

Definition 2.2.1 An $(H^{(n)}, G)$-automaton $\boldsymbol{A}$ is said to be *regular* if $G \approx G(\boldsymbol{A})$ holds.

We will discuss later the regularities of generalized group-matrix type automata.

2.3 Representations of general automata

In this section, we deal with representations of general automata by regular generalized group-matrix type automata.

Lemma 2.3.1 *Let* $\boldsymbol{A} = (\widehat{G}(H^{(n)}), X, \delta_\Psi)$ *be an* $(H^{(n)}, G)$*-automaton and* G' *be a group isomorphic to* G*. Then there exist a family of subgroups of* G', $H'^{(n)} = \{H'_1, H'_2, H'_3, \ldots, H'_n\}$*, satisfying the SP-condition, and an* $(H'^{(n)}, G')$*-automaton* $\boldsymbol{A}'$ *isomorphic to* $\boldsymbol{A}$*.*

Proof Let $H^{(n)} = \{H_1, H_2, H_3, \ldots, H_n\}$. From the fact that $G \approx G'$, there exists an isomorphism φ of G onto G'. Here, we put $H'_i = \varphi(H_i)$ for any $i = 1, 2, \ldots, n$. Then it is easily seen that $H'^{(n)} = \{H'_1, H'_2, H'_3, \ldots, H'_n\}$ satisfies the SP-condition.

Now, we define a set of inputs X' and a mapping Ψ' of X into $\widetilde{G'}(H'^{(n)})$ as follows:

(1) $X' = \{a' \mid a \in X\}$.

(2) $\Psi'(a') = \left(\widetilde{\varphi}(g_{pq})H'_q\right)$, if $\Psi(a) = (g_{pq}H_q)$.

Thus an $(H'^{(n)}, G')$-automaton $\boldsymbol{A}' = (\widehat{G'}(H'^{(n)}), X', \delta_{\Psi'})$ can be defined. It can be easily seen that, there exist an injective mapping ρ of $\widehat{G}(H^{(n)})$ onto $\widehat{G'}(H'^{(n)})$ and an injective mapping ξ of X onto X', such that $\rho(\delta_\Psi((g_pH_p), a)) = \delta_{\Psi'}(\rho((g_pH_p)), \xi(a))$ for any $(g_pH_p) \in \widehat{G}(H^{(n)})$ and any $a \in X$. For example, it is enough to put $\rho((g_pH_p)) = \left(\widetilde{\varphi}(g_p)H'_p\right)$ and $\xi(a) = a'$ for any $(g_pH_p) \in \widehat{G}(H^{(n)})$ and any $a \in X$.

This means that $\boldsymbol{A}'$ is isomorphic to $\boldsymbol{A}$.

Theorem 2.3.1 *Let* $\boldsymbol{A} = (S, X, \delta)$ *be an automaton and let* G *be a finite group such that* $G \approx G(\boldsymbol{A})$*. Then there exist a positive integer* n*, a family of subgroups of* G, $H^{(n)} = \{H_1, H_2, H_3, \ldots, H_n\}$ *satisfying the SP-condition and a regular* $(H^{(n)}, G)$*-automaton isomorphic to* $\boldsymbol{A}$*.*

Proof By Lemma 2.3.1, we can assume that $G = G(\boldsymbol{A})$ without loss of generality.

Let n be the cardinality of the set of all equivalence classes by the relation $\sim$ on S (for the relation $\sim$, see the proof of Theorem 1.3.1). Then we have $S = \bigcup_{i=1}^{n} S_i$ and $S_i \cap S_j = \emptyset, i \neq j$ where each $S_i = \{g(s_i) \mid g \in G(\boldsymbol{A})\}, i = 1, 2, \ldots, n$ is an equivalence class by the relation $\sim$ on S.

Now, we put $H_i = \{g \in G(\boldsymbol{A}) \mid g(s_i) = s_i\}$ for any $i = 1, 2, \ldots, n$.

First, we prove that each $H_i, i = 1, 2, \ldots, n$ is a subgroup of $G(\boldsymbol{A})$. To this end, since $G(\boldsymbol{A})$ is a finite group, it is enough to verify that $gh \in H_i$ holds for any $g, h \in H_i$. Let g and h be elements in H_i. Then by the definition of H_i, $g(s_i) = h(s_i) = s_i$ holds. Therefore, we have $gh(s_i) = g(s_i) = s_i$, i.e. $gh \in H_i$.

Next, we prove that $H^{(n)} = \{H_1, H_2, H_3, \ldots, H_n\}$ satisfies the SP-condition.

Let $h \in \bigcap_{g \in G(A)} (\bigcap_{i=1}^{n} gH_ig^{-1})$, i.e. $h \in gH_ig^{-1}$ for any $g \in G(\boldsymbol{A})$ and any $i = 1, 2, \ldots, n$. Notice that, for any $s \in S$, there exist some $i = 1, 2, \ldots, n$ and $g_i \in G(\boldsymbol{A})$ such that $s = g_i(s_i)$. For this g_i, we have $h \in g_iH_ig_i^{-1}$, namely $hg_i \in g_iH_i$. This implies $hg_i(s_i) = g_i(s_i)$. Since $s = g_i(s_i)$, we have $h(s) = s$. This means that $h = e$ holds. Thus we have $\bigcap_{g \in G(A)} (\bigcap_{i=1}^{n} gH_ig^{-1}) = \{e\}$.

Now, notice that, for any $i = 1, 2, \ldots, n$ and any $a \in X$, we can find a unique integer $k = 1, 2, \ldots, n$ and an element $h \in G(\boldsymbol{A})$ such that $\delta(s_i, a) = h(s_k)$. Hence we can define a mapping Ψ of X into $\widetilde{G(\boldsymbol{A})}(H^{(n)})$ as follows:

$$\Psi(a) = (g_{pq}H_q) \text{ where } g_{ij} = h \text{ if } j = k, \text{ and } g_{ij} = 0 \text{ if } j \neq k.$$

Adequacy of the definition Let $\delta(s_i, a) = g_{ij}(s_j) = g'_{ij}(s_j)$ where $g_{ij}, g'_{ij} \in G(\boldsymbol{A})$. From the above, $g_{ij}^{-1}g'_{ij}(s_j) = s_j$ holds. By the definition of H_j, we have $g_{ij}^{-1}g'_{ij} \in H_j$, namely $g_{ij}H_j = g'_{ij}H_j$.

Proof of $\Psi(a) \in \widetilde{G(\boldsymbol{A})}(H^{(n)})$ It is obvious that, for any $i = 1, 2, \ldots, n$, there exists a unique integer $j = 1, 2, \ldots, n$ such that $g_{ij} \neq 0$.

Therefore, we will prove only that, if $g_{ij} \neq 0$, then $H_i \subseteq g_{ij}H_jg_{ij}^{-1}$. Since $g_{ij} \neq 0$, we have $\delta(s_i, a) = g_{ij}(s_j)$. Let g be an element in H_i. Then $g(s_i) = s_i$ holds. Thus we have $g_{ij}(s_j) = \delta(s_i, a) = \delta(g(s_i), a) = g(\delta(s_i, a)) = gg_{ij}(s_j)$. Cosequently, $gg_{ij} \in g_{ij}H_j$ holds and hence we have $g \in g_{ij}H_jg_{ij}^{-1}$. This means that $H_i \subseteq g_{ij}H_jg_{ij}^{-1}$ holds.

Thus we have an $(H^{(n)}, G(\boldsymbol{A}))$-automaton $\boldsymbol{A}' = (\widehat{G(\boldsymbol{A})}(H^{(n)}), X, \delta_\Psi)$.

Now, we will prove that $\boldsymbol{A} \approx \boldsymbol{A}'$. To this end, we define first ξ as the identity mapping on X, i.e. $\xi(a) = a$ for any $a \in X$. Next, we define the mapping ρ as follows:

$\rho(s) = (0, \ldots, 0, g_iH_i, 0, \ldots, 0)$ for $s \in S$ where i and g_i are an integer and an element in $G(\boldsymbol{A})$, respectively, determined with respect to s such that $s = g_i(s_i)$.

Adequacy of the definition Assume that $s = g_i(s_i) = g_i'(s_i)$ holds for $g_i, g_i' \in G(\boldsymbol{A})$. Then it is directly shown that $g_iH_i = g_i'H_i$ holds.

Moreover, it is obvious that ρ is a mapping of S onto $\widehat{G(\boldsymbol{A})}(H^{(n)})$.

Now, we prove that ρ is a surjective mapping. Suppose that $\rho(s) = \rho(t)$ holds where $s, t \in S$. Then there exist an integer $i = 1, 2, \ldots, n$ and elements $g_i, h_i \in G(\boldsymbol{A})$ such that $s = g_i(s_i)$, $t = h_i(s_i)$ and $g_iH_i = h_iH_i$. By $g_iH_i = h_iH_i$, we have $g_i(s_i) = h_i(s_i)$, and hence $s = t$. This means that ρ is a surjective mapping.

Finally, we show that $\rho(\delta(s, a)) = \delta_\Psi(\rho(s), \xi(a))$ holds for any $s \in S$ and any $a \in X$. Suppose that $s = g_i(s_i)$ and $\delta(s_i, a) = g_{ij}(s_j)$ where $g_i, g_{ij} \in G(\boldsymbol{A})$. Then we have $\rho(\delta(s, a)) = \rho(\delta(g_i(s_i), a)) = \rho(g_i(\delta(s_i, a))) = \rho(g_ig_{ij}(s_j)) = (0, \ldots, 0, g_ig_{ij}H_j, 0, \ldots, 0)$. On the other hand, we have:

$$
\begin{aligned}
\delta_\Psi(\rho(s), a) &= \rho(s)\Psi(a) \\
&= (0, \ldots, 0, g_iH_i, 0, \ldots, 0) \begin{pmatrix} & * & & \vdots & & * & \\ 0 & \cdots & 0 & g_{ij}H_j & 0 & \cdots & 0 \\ & * & & \vdots & & * & \end{pmatrix} \begin{matrix} \\ i\text{-th} \\ \\ \end{matrix} \\
&\qquad\qquad\qquad\qquad\qquad\qquad\qquad\qquad \text{(column } j\text{-th)} \\
&= (0, \ldots, 0, (g_iH_i)(g_{ij}H_j), 0, \ldots, 0) = (0, \ldots, 0, g_ig_{ij}H_i, 0, \ldots, 0).
\end{aligned}
$$

Thus we have $\rho(\delta(s,a)) = \delta_\Psi(\rho(s),a) = \delta_\Psi(\rho(s),\xi(a))$. Hence $\boldsymbol{A} \approx \boldsymbol{A}'$ holds.

To conclude, we prove the regularity of $\boldsymbol{A}'$. Since $\boldsymbol{A}' \approx \boldsymbol{A}$, we have $G(\boldsymbol{A}') \approx G(\boldsymbol{A})$. This means that $\boldsymbol{A}'$ is regular.

Example 2.3.1 Fig. 2.1 describes the state transition diagram of an automaton $\boldsymbol{A} = (S, X, \delta)$. We will give a representation of $\boldsymbol{A}$ by a regular generalized group-matrix type automaton.

$$G(\boldsymbol{A}) = \{e, (p\,q), (s\,t), (p\,q)(s\,t)\} \text{ where } e \text{ is the identity of } G(\boldsymbol{A}).$$

$$S = \{p, q, r, s, t\} = \{p, q\} \cup \{r\} \cup \{s, t\}\,.$$

$$S_1 = \{p, q\}\,,\ s_1 = p.\ \ S_2 = \{r\}\,,\ s_2 = r.\ \ S_3 = \{s, t\}\,,\ s_3 = s.$$

$$H_1 = \{e, (s\,t)\}\,,\ \ H_2 = G(\boldsymbol{A}),\ \ H_3 = \{e, (p\,q)\}\,.$$

$$H^{(3)} = \{H_1, H_2, H_3\}\,.\ \ G(\boldsymbol{A}) = H_1 + (p\,q)H_1 = H_2 = H_3 + (s\,t)H_3.$$

$$\widehat{G(\boldsymbol{A})}(H^{(3)}) = \{(H_1,0,0),\ ((p\,q)H_1,0,0),\ (0,H_2,0),\ (0,0,H_3),(0,0,(s\,t)H_3)\}.$$

$$\delta(s_1,a) = \delta(p,a) = r = e(s_2),\ \delta(s_2,a) = \delta(r,a) = r = e(s_2),$$

$$\delta(s_3,a) = \delta(s,a) = s = e(s_3),\ \delta(s_1,b) = \delta(p,b) = p = e(s_1),$$

$$\delta(s_2,b) = \delta(r,b) = r = e(s_2),\ \delta(s_3,b) = \delta(s,b) = t = (s\,t)(s_3).$$

$$\Psi(a) = \begin{pmatrix} 0 & H_2 & 0 \\ 0 & H_2 & 0 \\ 0 & 0 & H_3 \end{pmatrix},\quad \Psi(b) = \begin{pmatrix} H_1 & 0 & 0 \\ 0 & H_2 & 0 \\ 0 & 0 & (s\,t)H_3 \end{pmatrix}.$$

$$\boldsymbol{A}' = \left(\widehat{G(\boldsymbol{A})}(H^{(3)}), \{a,b\}\,, \delta_\Psi\right) \approx \boldsymbol{A}.$$

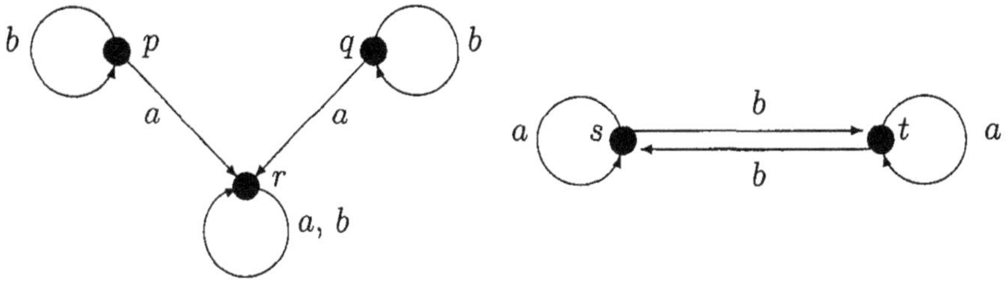

Figure 2.1: State transition diagram of $\boldsymbol{A}$

2.4 Regularities

In the present section, we will investigate the regularities of generalized group-matrix type automata. However, we will deal with only a very special case for a generalized group-matrix type automaton of order 1.

First, let $K = \{k \in G \mid \exists x \in X^*, \delta_\Psi(H, x) = kH\}$ for an $(H^{(1)}, G)$-automaton $\boldsymbol{A} = (\widehat{G}(H^{(1)}), X, \delta_\Psi)$ where $H^{(1)} = \{H\}$.

Notice that $H \subseteq K$ holds and that K forms a subgroup of G. Then we have the following result.

Theorem 2.4.1 *If K satisfies the following two conditions, then $\boldsymbol{A}$ is not a regular.*

(1) $H \neq K$.

(2) $\bigcap_{g \in G} gKg^{-1} = \{e\}$ *where e is the identity of G.*

Proof Let $f \in K \setminus H$. We consider the following mapping ρ of $\widehat{G}(H^{(1)})$ into itself: $\rho(gH) = fgH$ if $g \in K$, and $\rho(hH) = hH$ if $h \notin K$.

Then $\rho \in G(\boldsymbol{A})$ and $\rho \neq \rho_e$. Now, suppose that $\boldsymbol{A}$ is regular. Then there exists $d \in G$ such that $\rho = \rho_d$. Thus we have $\rho(gH) = fgH = \rho_d(gH) = dgH$ for any $g \in K$. Therefore, we have $d \in fgHg^{-1} \subseteq K$. On the other hand, we have $\rho(hH) = hH =$

$\rho_d(hH) = dhH$ for any $h \notin K$, and hence $d \in \bigcap_{h \in G \setminus K} hKh^{-1}$. Thus $d \in \bigcap_{g \in G} gKg^{-1}$ holds. Hence $d = e$ must be satisfied. However, this contradicts the fact that $\rho \neq \rho_e$.

Thus $\boldsymbol{A}$ is not regular.

Remark 2.4.1 In connection with the condition (1), consider the following: It is easy to deal with an $(H^{(1)}, G)$-automaton with $H = K$. For this automaton, we obtain easily $G(\boldsymbol{A}) \approx S(|G| \,/\, |H|)$ where $S(p)$ denotes the symmetric group of degree p.

In particular, we have $G \approx S(|G| \,/\, |H|)$ if $\boldsymbol{A}$ is regular. Therefore, $|G| = (|G| \,/\, |H|)!$ holds. This indicates a very special case.

Remark 2.4.2 Strictly speaking, the condition (2) can be replaced by

$$K \bigcap (\bigcap_{g \in G \setminus K} gHg^{-1}) = \{e\}$$

as shown in the proof of Theorem 2.4.1. However, it seems that the condition (2) is enough from the practical point of view. For this, see the following corollary.

Corollary 2.4.1 *Let G be a simple group. Then $\boldsymbol{A}$ is a $(1, G)$-automaton if an $(H^{(1)}, G)$-automaton $\boldsymbol{A}$ is regular.*

Proof In the case $|G| \leq 2$, by Proposition 2.2.1, the assertion of the corollary holds true. Thus we assume that $|G| > 2$.

When $K = G$, $\boldsymbol{A}$ is strongly connected. Then by Proposition 2.2.2, $\boldsymbol{A}$ is a $(1, G)$-automaton.

Now, let $\boldsymbol{A}$ not be a $(1, G)$-automaton and $K \neq G$. Then $H \neq \{e\}$ holds where $H^{(1)} = \{H\}$. Suppose that $H = K$. By Remark 2.4.1, G is a symmetric group of degree $|G|$. However, this contradicts the fact that G is simple. Consequently, $H \neq K$ holds. By $K \neq G$, the normal subgroup $\bigcap_{g \in G} gKg^{-1}$ $(\neq G)$ of G must be equal to $\{e\}$. Thus by Theorem 2.4.1, $\boldsymbol{A}$ is not regular. This is a contradiction, and hence $\boldsymbol{A}$ is a $(1, G)$-automaton.

Chapter 3

Classes of Automata as Posets

In this chapter, we will discuss a few classes of automata as partially ordered sets (posets) whose orders are induced by homomorphism relation. For some classes of automata called *strong classes*, we will provide a systematic method to compute the minimal elements in the set of all upper bounds for any given two elements. Moreover, for a class which forms a lattice, we provide a method to obtain the greatest lower bound of any given two elements as well. In both methods, the direct product of automata plays an important role. This means that the direct product of automata is not only a formal definition but also contains much useful information.

The contents of this chapter are based on the results in [36]. A category-theoretic and a more general treatment can be also seen in [37] and [42]. The contents of this chapter represent a kind of decomposition theory of automata. As for a standard decomposition theory of automata, i.e. Krohn-Rhodes theory, see e.g. [16] and [20].

3.1 Introductory notions and results

In this chapter, almost all notions of which we need to develop the contents have been already introduced, but we will mention them and their related results again.

Definition 3.1.1 An *automaton* (more exactly an X*-automaton*) $\boldsymbol{A}$, denoted by $\boldsymbol{A} = (S, X)$, consists of the following data: (i) S is a finite nonempty set of states. (ii) X is a finite nonempty set of inputs. (iii) There exists a function δ_A of $S \times X^*$ into S, called a *state transition function* such that $\delta_A(s, uv) = \delta_A(\delta_A(s, u), v)$ and $\delta_A(s, \epsilon) = s$ for any $s \in S$ and any $u, v \in X^*$ where X^* is the free monoid generated by X and ϵ is its identity, i.e. the *empty word*.

In what follows, su^A will be used to denote $\delta_A(s, u)$.

Definition 3.1.2 Let $\boldsymbol{A} = (S, X)$ and $\boldsymbol{B} = (T, X)$ be two automata. If $T \subseteq S$ and $tu^B = tu^A$ for any $(t, u) \in T \times X^*$, then $\boldsymbol{B}$ is called a *subautomaton* of $\boldsymbol{A}$.

Definition 3.1.3 An automaton $\boldsymbol{A} = (S, X)$ is said to be *cyclic*, if there is a state $s \in S$ satisfying the following condition: For any $t \in S$ there exists an $u \in X^*$ such that $su^A = t$.

In the above definition, s is called a *generator* of $\boldsymbol{A}$ and the set of all generators of $\boldsymbol{A}$ is denoted by $\mathrm{Gen}(\boldsymbol{A})$. Moreover, if $S = \mathrm{Gen}(\boldsymbol{A})$, $\boldsymbol{A}$ is said to be *strongly connected.*

Definition 3.1.4 Let $\boldsymbol{A} = (S, X)$ and $\boldsymbol{B} = (T, X)$ be two automata. Then the automaton $\boldsymbol{A} \times \boldsymbol{B} = (S \times T, X)$ is called the *direct product* of $\boldsymbol{A}$ and $\boldsymbol{B}$. Here the state transition function $\delta_{A\times B}$ is defined as $(s, t)a^{A\times B} = (sa^A, ta^B)$ for any $(s, t) \in S \times T$ and all $a \in X$. Two automata $\boldsymbol{A}$ and $\boldsymbol{B}$, whose direct product $\boldsymbol{A} \times \boldsymbol{B}$ is strongly connected, are said to be *strongly related.* For a condition such that two automata may be strongly related, see [21].

Definition 3.1.5 Let $\boldsymbol{A} = (S, X)$ and $\boldsymbol{B} = (T, X)$ be two automata. If there exists a mapping ρ of S onto T such that for any $(s, a) \in S \times X$ we have $\rho(sa^A) = \rho(s)a^B$, then ρ is called a *homomorphism* of $\boldsymbol{A}$ onto $\boldsymbol{B}$ and $\boldsymbol{B}$ is called a *homomorphic image* of $\boldsymbol{A}$. We denote the set of all homomorphisms of $\boldsymbol{A}$ onto $\boldsymbol{B}$ by $\mathrm{Hom}(\boldsymbol{A}, \boldsymbol{B})$. Moreover, if ρ is a bijection, then ρ is called an *isomorphism* of $\boldsymbol{A}$ onto $\boldsymbol{B}$ and $\boldsymbol{A}$ is said to be *isomorphic* to $\boldsymbol{B}$. The set of all isomorphisms of $\boldsymbol{A}$ onto $\boldsymbol{B}$ is denoted by $\mathrm{Iso}(\boldsymbol{A}, \boldsymbol{B})$.

Definition 3.1.6 In the above definition, if $\boldsymbol{A} = \boldsymbol{B}$, then we denote $\mathrm{Iso}(\boldsymbol{A}, \boldsymbol{A})$ by $\mathrm{Aut}(\boldsymbol{A})$ and we call an element of $\mathrm{Aut}(\boldsymbol{A})$ an *automorphism* of $\boldsymbol{A}$. $\mathrm{Aut}(\boldsymbol{A})$ forms a group with mapping composition and is called the *automorphism group* of $\boldsymbol{A}$.

Definition 3.1.7 An automaton $\boldsymbol{A} = (S, X)$ is said to be *transitive* if the following condition is satisfied: For any $(s, t) \in S \times S$, there exists $\rho \in \mathrm{Aut}(\boldsymbol{A})$ such that $t = \rho(s)$. Moreover, a transitive automaton is said to be *quasiperfect*, if it is strongly connected.

Definition 3.1.8 Let $A = (G, X)$ be an automaton satisfying the following conditions: (i) G is a finite group. (ii) There exists a mapping Ψ of X into G such that $\Psi(X) = \{\Psi(a) \mid a \in X\}$ is a generating set of G. (iii) The state transition function δ_A is defined as $ga^A = g\Psi(a)$ for any $(g, a) \in G \times X$. Then an automaton $\boldsymbol{A} = (G, X)$ is called a *group-type automaton* (more exactly, a *group-type automaton with group* G) (or $(1, G)$*-automaton*).

Theorem 3.1.1 *A group-type automaton* $\boldsymbol{A} = (G, X)$ *is quasiperfect and* $\mathrm{Aut}(\boldsymbol{A}) \approx G$ *holds where* $\approx$ *denotes an isomorphism relation between two groups. Conversely, a quasiperfect automaton* $\boldsymbol{A} = (S, X)$ *is isomorphic to some group-type automaton* $\boldsymbol{A}' = (G, X)$ *such that* $G \approx \mathrm{Aut}(\boldsymbol{A})$.

Definition 3.1.9 An automaton $\boldsymbol{A} = (S, X)$ is said to be *abelian* (or *commutative*) if the following condition is satisfied: For any $s \in S$ and all $a, b \in X$, we have $s(ab)^A = s(ba)^A$. Moreover, an abelian automaton is said to be *perfect* if it is strongly connected.

Theorem 3.1.2 *A perfect automaton* $\boldsymbol{A} = (S, X)$ *is quasiperfect and* $\mathrm{Aut}(\boldsymbol{A})$ *is an abelian group. Consequently, a perfect automaton is isomorphic to some group-type automaton with an abelian group. Conversely, a group-type automaton with an abelian group is a perfect automaton.*

3.2 Classes of automata

We consider first the set of all automata as a poset where the partial order is given by homomorphism relation.

Definition 3.2.1 Let $\boldsymbol{A} = (S, X)$ and $\boldsymbol{B} = (T, X)$ be two automata. Then $\boldsymbol{A} \leq \boldsymbol{B}$ means that $\boldsymbol{A}$ is a homomorphic image of $\boldsymbol{B}$. Moreover, we identify $\boldsymbol{A}$ with $\boldsymbol{B}$, denoted by $\boldsymbol{A} = \boldsymbol{B}$, if $\boldsymbol{A}$ is isomorphic to $\boldsymbol{B}$. By this definition, the set of all X-automata forms a partially ordered set (poset) in which the order is given by $\leq$. In what follows, we denote this set by $V_X(Au)$, or simply by $V(Au)$.

We will deal with subposets of $V(Au)$.

Definition 3.2.2 We denote the set of all X-automata (notice that isomorphic automata are identified) having a property P by $V_X(P)$ (or simply $V(P)$) and we call it a *class*. Then $V_X(P)$ is considered as a subposet of $V_X(Au)$. Moreover, if the property P is preserved by homomorphism relation (i.e. if $\boldsymbol{A} \in V(P)$ and $\mathrm{Hom}(\boldsymbol{A}, \boldsymbol{A}') \neq \emptyset$, then $\boldsymbol{A}' \in V(P)$), $V(P)$ is called a *strong class*.

Definition 3.2.3 Let $V(P)$ be a class and let $\boldsymbol{A}, \boldsymbol{B} \in V(P)$. Then $[\boldsymbol{A} \circ \boldsymbol{B}]_{V(P)}$ denotes the set of all minimal elements in the set of all upper bounds of the set $\{\boldsymbol{A}, \boldsymbol{B}\}$ in $V(P)$. Namely, $\boldsymbol{C} \in [\boldsymbol{A} \circ \boldsymbol{B}]_{V(P)}$ means the following: $\boldsymbol{C} \in V(P)$, $\boldsymbol{C} \geq \boldsymbol{A}$ and $\boldsymbol{C} \geq \boldsymbol{B}$, and for any $\boldsymbol{D} \in V(P)$ such that $\boldsymbol{D} \geq \boldsymbol{A}$, $\boldsymbol{D} \geq \boldsymbol{B}$ and $\boldsymbol{C} \geq \boldsymbol{D}$, we have $\boldsymbol{C} = \boldsymbol{D}$. Dually, we can define the set of all maximal elements in the set of all lower bounds of the set $\{\boldsymbol{A}, \boldsymbol{B}\}$ in $V(P)$ and we denote it by $[\boldsymbol{A} * \boldsymbol{B}]_{V(P)}$.

We will try to compute $[\boldsymbol{A} \circ \boldsymbol{B}]_{V(P)}$ for any $\boldsymbol{A}, \boldsymbol{B} \in V(P)$.

Definition 3.2.4 Let $\boldsymbol{A} = (S, X), \boldsymbol{B} = (T, X)$ and $\boldsymbol{C} = (U, X)$ be automata. Assume that $g \in \mathrm{Hom}(\boldsymbol{C}, \boldsymbol{A})$ and $h \in \mathrm{Hom}(\boldsymbol{C}, \boldsymbol{B})$. Then, by $g\times h(\boldsymbol{C})$, we denote the automaton $g\times h(\boldsymbol{C}) = (g\times h(U), X)$ where $g \times h(U) = \{(g(u), h(u)) \mid u \in U\}$ and the state transition function $\delta_{g\times h(C)}$ is defined as $(g(u), h(u))a^{g\times h(C)} = (g(u)a^A, h(u)a^B)$ for any $(u, a) \in U \times X$.

Remark 3.2.1 Notice that, in the above definition, $g \times h(\boldsymbol{C})$ is well defined as an automaton and it is a subautomaton of $\boldsymbol{A} \times \boldsymbol{B}$. In particular, if $\boldsymbol{A}$ and $\boldsymbol{B}$ are strongly related, then $g \times h(\boldsymbol{C}) = \boldsymbol{A} \times \boldsymbol{B}$.

The following result is obvious.

Lemma 3.2.1 $\boldsymbol{C} \geq g \times h(\boldsymbol{C}) \geq \boldsymbol{A}, \boldsymbol{B}$.

For a strong class, we have the following result.

Lemma 3.2.2 *Let $V(P)$ be a strong class. Assume $\boldsymbol{C} \in [\boldsymbol{A} \circ \boldsymbol{B}]_{V(P)}$, $g \in \mathrm{Hom}(\boldsymbol{C}, \boldsymbol{A})$ and $h \in \mathrm{Hom}(\boldsymbol{C}, \boldsymbol{B})$. Then we have $\boldsymbol{C} = g \times h(\boldsymbol{C})$.*

Proof By Lemma 3.2.1, we have $\boldsymbol{C} \geq g \times h(\boldsymbol{C}) \geq \boldsymbol{A}, \boldsymbol{B}$. Since $V(P)$ is a strong class, we have $g \times h(\boldsymbol{C}) \in V(P)$. Therefore, by the definition of $[\boldsymbol{A} \circ \boldsymbol{B}]_{V(P)}$, we have $\boldsymbol{C} = g \times h(\boldsymbol{C})$.

By Remark 3.2.1, the following result can be obtained.

Proposition 3.2.1 *Let $V(P)$ be a strong class and let $\boldsymbol{A}, \boldsymbol{B} \in V(P)$ be strongly related automata. Then $[\boldsymbol{A} \circ \boldsymbol{B}]_{V(P)} \subseteq \{\boldsymbol{A} \times \boldsymbol{B}\}$.*

We will give a method to find elements of $[\boldsymbol{A} \circ \boldsymbol{B}]_{V(P)}$ for any $\boldsymbol{A}, \boldsymbol{B} \in V(P)$ when $V(P)$ is a strong class.

Theorem 3.2.1 *Let $V(P)$ be a strong class. Assume $\boldsymbol{A} = (S, X)$ and $\boldsymbol{B} = (T, X)$ are elements of $V(P)$ and that $\boldsymbol{C} = (U, X)$ is an element of $[\boldsymbol{A} \circ \boldsymbol{B}]_{V(P)}$. Then $\boldsymbol{C}$ is a subautomaton of $\boldsymbol{A} \times \boldsymbol{B}$. Moreover, let π_A and π_B be the mappings such that $\pi_A((s,t)) = s$ and $\pi_B((s,t)) = t$ for any $(s,t) \in S \times T$. Then we have $\pi_A(U) = \{\pi_A(u) \mid u \in U\} = S$ and $\pi_B(U) = \{\pi_B(u) \mid u \in U\} = T$. Next, let $\boldsymbol{C}' = (U', X) \in V(P)$ be a subautomaton of $\boldsymbol{A} \times \boldsymbol{B}$ such that $\pi_A(U') = S$, $\pi_B(U') = T$ and $|U'| = \min\{|U| \mid \boldsymbol{C} = (U, X) \in V(P)$ is a subautomaton of $\boldsymbol{A} \times \boldsymbol{B}$ such that $\pi_A(U) = S$ and $\pi_B(U) = T\}$ where $|K|$ denotes the cardinality of the set K. Then we have $\boldsymbol{C}' \in [\boldsymbol{A} \circ \boldsymbol{B}]_{V(P)}$.*

Proof of the First Part Let $\boldsymbol{C} \in [\boldsymbol{A} \circ \boldsymbol{B}]_{V(P)}$. Then there exist $g \in \mathrm{Hom}(\boldsymbol{C}, \boldsymbol{A})$ and $h \in \mathrm{Hom}(\boldsymbol{C}, \boldsymbol{B})$. By Lemma 3.2.2, we have $\boldsymbol{C} = g \times h(\boldsymbol{C})$. Thus, by Remark 3.2.1, $\boldsymbol{C}$ is a subautomaton of $\boldsymbol{A} \times \boldsymbol{B}$. Moreover, by the fact that g and h are surjections, we have $\pi_A(U) = S$, and $\pi_B(U) = T$.

Proof of the Second Part Let $\boldsymbol{C}'$ be an automaton satisfying the condition in Theorem 3.2.1. Then it can be proved that $\boldsymbol{C}' \geq \boldsymbol{A}, \boldsymbol{B}$.

Now, suppose $\boldsymbol{C}' \notin [\boldsymbol{A} \circ \boldsymbol{B}]_{V(P)}$. Since $\boldsymbol{C}' \geq \boldsymbol{A}, \boldsymbol{B}$ and $\boldsymbol{C}' \in V(P)$, there exists an automaton $\boldsymbol{C} = (U, X)$ such that $\boldsymbol{C}' \geq \boldsymbol{C}$ and $\boldsymbol{C} \in [\boldsymbol{A} \circ \boldsymbol{B}]_{V(P)}$. By the proof of the first part of the theorem, we can see that $\boldsymbol{C} = (U, X) \in V(P)$ is a subautomaton of $\boldsymbol{A} \times \boldsymbol{B}$ such that $\pi_A(U) = S$ and $\pi_B(U) = T$. On the other hand, since $\boldsymbol{C}' \geq \boldsymbol{C}$ and $\boldsymbol{C}' \neq \boldsymbol{C}$, we have $|U'| > |U|$. However, this is a contradiction. Thus, we have $\boldsymbol{C}' \in [\boldsymbol{A} \circ \boldsymbol{B}]_{V(P)}$.

Let $V_X(Au)$, $V_X(C)$ and $V_X(S)$ be the classes of all automata, cyclic automata and strongly connected automata, respectively. Then all these classes are strong classes. For these classes, we have the following results.

Corollary 3.2.1 *For any $\boldsymbol{A}, \boldsymbol{B} \in V(Au)$, we have $[\boldsymbol{A} \circ \boldsymbol{B}]_{V(Au)} \neq \emptyset$ and $[\boldsymbol{A} \circ \boldsymbol{B}]_{V(Au)} \subsetneq \{\boldsymbol{C} \mid \boldsymbol{C} = (U, X)$ is a subautomaton of $\boldsymbol{A} \times \boldsymbol{B}$ such that $\pi_A(U) = S$ and $\pi_B(U) = T\}$. Moreover, let $\boldsymbol{C}' = (U', X)$ be a subautomaton of $\boldsymbol{A} \times \boldsymbol{B}$ such that $\pi_A(U') = S$, $\pi_B(U') = T$ and $|U'| = \min\{|U| \mid \boldsymbol{C} = (U, X)$ is a subautomaton of $\boldsymbol{A} \times \boldsymbol{B}$ such that $\pi_A(U) = S$ and $\pi_B(U) = T\}$. Then we have $\boldsymbol{C}' \in [\boldsymbol{A} \circ \boldsymbol{B}]_{V(Au)}$.*

Proof Notice that $\boldsymbol{A} \times \boldsymbol{B} \in V(Au)$ and $\boldsymbol{A} \times \boldsymbol{B} \geq \boldsymbol{A}, \boldsymbol{B}$. Thus we have $[\boldsymbol{A} \circ \boldsymbol{B}]_{V(Au)} \neq \emptyset$. The rest of the corollary can be proved by Theorem 3.2.1.

Corollary 3.2.2 *For any $\boldsymbol{A}, \boldsymbol{B} \in V(C)$, we have $[\boldsymbol{A} \circ \boldsymbol{B}]_{V(C)} \neq \emptyset$ and $[\boldsymbol{A} \circ \boldsymbol{B}]_{V(C)} \subsetneq \{\boldsymbol{C} \mid \boldsymbol{C} = (U, X)$ is a cyclic subautomaton of $\boldsymbol{A} \times \boldsymbol{B}$ such that $\pi_A(U) = S$ and $\pi_B(U) = T\}$. Moreover, let $\boldsymbol{C}' = (U', X)$ be a cyclic subautomaton of $\boldsymbol{A} \times \boldsymbol{B}$ such that $\pi_A(U') = S$, $\pi_B(U') = T$ and $|U'| = \min\{|U| \mid \boldsymbol{C} = (U, X)$ is a cyclic subautomaton of $\boldsymbol{A} \times \boldsymbol{B}$ such that $\pi_A(U) = S$ and $\pi_B(U) = T\}$. Then we have $\boldsymbol{C}' \in [\boldsymbol{A} \circ \boldsymbol{B}]_{V(C)}$.*

Proof Let $(s, t) \in \mathrm{Gen}(\boldsymbol{A}) \times \mathrm{Gen}(\boldsymbol{B})$ and $U[s, t] = \{(su^A, tu^B) \mid u \in X^*\}$. Then we have the automaton $\boldsymbol{C}[s, t] = (U[s, t], X)$ such that the state transition function $\delta_{C[s,t]}$ is defined as $(s', t')a^{C[s,t]} = (s'a^A, t'a^B)$ for any $(s', t') \in U[s, t]$ and all $a \in X$. Then it can be proved that $\boldsymbol{C}[s, t]$ is a cyclic subautomaton of $\boldsymbol{A} \times \boldsymbol{B}$ such that $\pi_A(U[s, t]) = S$, $\pi_B(U[s, t]) = T$ and $\boldsymbol{C}[s, t] \geq \boldsymbol{A}, \boldsymbol{B}$. Thus, we have $[\boldsymbol{A} \circ \boldsymbol{B}]_{V(C)} \neq \emptyset$. The rest of the corollary can be proved by Theorem 3.2.1.

Notice that we have the following: $\{\boldsymbol{C} \mid \boldsymbol{C} = (U, X)$ is a cyclic subautomaton of $\boldsymbol{A} \times \boldsymbol{B}$ such that $\pi_A(U) = S$ and $\pi_B(U) = T\} = \{\boldsymbol{C}[s,t] \mid (s,t) \in \mathrm{Gen}(\boldsymbol{A}) \times \mathrm{Gen}(\boldsymbol{B})\}$.

Corollary 3.2.3 *For any $\boldsymbol{A}, \boldsymbol{B} \in V(S)$, we have $[\boldsymbol{A} \circ \boldsymbol{B}]_{V(S)} \neq \emptyset$ and $[\boldsymbol{A} \circ \boldsymbol{B}]_{V(S)} \subseteq \{\boldsymbol{C} \mid \boldsymbol{C}$ is a strongly connected subautomaton of $\boldsymbol{A} \times \boldsymbol{B}\}$. Moreover, let $\boldsymbol{C}' = (U', X)$ be a strongly connected subautomaton of $\boldsymbol{A} \times \boldsymbol{B}$ such that $|U'| = \min\{|U| \mid \boldsymbol{C} = (U, X)$ is a strongly connected subautomaton of $\boldsymbol{A} \times \boldsymbol{B}\}$. Then we have $\boldsymbol{C}' \in [\boldsymbol{A} \circ \boldsymbol{B}]_{V(S)}$.*

Proof Let $\boldsymbol{C} = (U, X)$ be a subautomaton of $\boldsymbol{A} \times \boldsymbol{B}$. Since $\boldsymbol{A}$ and $\boldsymbol{B}$ are strongly connected, $\pi_A(U) = S$, $\pi_B(U) = T$ and $\boldsymbol{C} \geq \boldsymbol{A}, \boldsymbol{B}$. Thus we have $[\boldsymbol{A} \circ \boldsymbol{B}]_{V(S)} \neq \emptyset$. The rest of the corollary can be proved by Theorem 3.2.1.

Now, we deal with the relationship among classes.

Proposition 3.2.2 *Let $V(P)$ be a strong class and let $V(Q)$ be a class. Moreover, assume that $V(P) \subseteq V(Q)$. Then for any $\boldsymbol{A}, \boldsymbol{B} \in V(P)$, we have $[\boldsymbol{A} \circ \boldsymbol{B}]_{V(P)} \subseteq [\boldsymbol{A} \circ \boldsymbol{B}]_{V(Q)}$.*

Proof Assume $\boldsymbol{C} \in [\boldsymbol{A} \circ \boldsymbol{B}]_{V(P)}$. By $\boldsymbol{C} \in V(P)$, we have $\boldsymbol{C} \in V(Q)$. Moreover, by $\boldsymbol{C} \geq \boldsymbol{A}, \boldsymbol{B}$, there exists some $\boldsymbol{D} \in V(Q)$ such that $\boldsymbol{C} \geq \boldsymbol{D} \geq \boldsymbol{A}, \boldsymbol{B}$ and $\boldsymbol{D} \in [\boldsymbol{A} \circ \boldsymbol{B}]_{V(Q)}$. Since $V(P)$ is a strong class, we have $\boldsymbol{D} \in V(P)$. Therefore, by the definition of $[\boldsymbol{A} \circ \boldsymbol{B}]_{V(P)}$, we have $\boldsymbol{C} = \boldsymbol{D}$. Consequently, we have $\boldsymbol{C} \in [\boldsymbol{A} \circ \boldsymbol{B}]_{V(Q)}$, i.e. $[\boldsymbol{A} \circ \boldsymbol{B}]_{V(P)} \subseteq [\boldsymbol{A} \circ \boldsymbol{B}]_{V(Q)}$.

3.3 Classes which do not form lattices

In this section, we will provide some classes which do not form lattices. First, we prove that $V_X(S)$ $(|X| \geq 2)$ does not form a lattice though $V_X(S)$ forms a lattice for a singleton set X.

Let $\boldsymbol{A} = (G, X)$ be a group-type automaton with group G having a generating set $\{g_1, g_2, \ldots, g_n\}$ where $X = \{a_1, a_2, \ldots, a_n\}$ and the state transition function δ_A is defined as $ga_i^A = gg_i$ for any $g \in G$ and any $i = 1, 2, \ldots, n$. Moreover, let H be a subgroup of G. Then the automaton $\boldsymbol{A}/H = (G/H, X)$ where $G/H = \{Hg \mid g \in G\}$ and the state transition function $\delta_{A/H}$ is defined as $Hga_i^{A/H} = Hgg_i$ for any $g \in G$ and any $i = 1, 2, \ldots, n$, is given and we have $\boldsymbol{A}/H \in V(S)$ and $\boldsymbol{A} \geq \boldsymbol{A}/H$.

Lemma 3.3.1 *Let H and K be subgroups of G. If $[\boldsymbol{A}/H \circ \boldsymbol{A}/K]_{V(S)}$ is a singleton set, then there exists $s \in G$ satisfying the following property: For any $r \in G$, there exists $\alpha \in G$ such that $H \cap r^{-1}Kr \subseteq \alpha^{-1}(H \cap s^{-1}Ks)\alpha$.*

Proof Let $\boldsymbol{A}/H \times \boldsymbol{A}/K = (G/H \times G/K, X)$. For any $(f, g) \in G \times G$, we put $S[f, g] = \{(Hfx, Kgx) \mid x \in G\} \subseteq G/H \times G/K$. Then we have the strongly connected subautomaton $[\boldsymbol{A}/H \times \boldsymbol{A}/K](f, g) = (S[f, g], X)$ of $\boldsymbol{A}/H \times \boldsymbol{A}/K$. Consequently, there exists some $(f', g') \in G \times G$ such that $[\boldsymbol{A}/H \times \boldsymbol{A}/K](f', g') = [\boldsymbol{A}/H \circ \boldsymbol{A}/K]_{V(S)}$. Thus, for any $(f, g) \in G \times G$, we have $\mathrm{Hom}([\boldsymbol{A}/H \times \boldsymbol{A}/K](f, g), [\boldsymbol{A}/H \times \boldsymbol{A}/K](f', g')) \neq \emptyset$. Now, we put $s = g'f'^{-1}$. Notice that for any $r \in G$ and all $x \in H \cap r^{-1}Kr$ we have $Hx = H$ and $Krx = Kr$. Since $\{g_1, g_2, \ldots, g_n\}$ is a generating set of G, there exists some $z \in X^*$ such that for any $(f, g) \in G \times G$ we have $(Hf, Kg)z^{A/H \times A/K} = (Hfx, Kgx)$. Therefore, we have $(H, Kr)z^{A/H \times A/K} = (Hx, Krx) = (H, Kr)$. Let ψ be an element of $\mathrm{Hom}([\boldsymbol{A}/H \times \boldsymbol{A}/K](e, r), [\boldsymbol{A}/H \times \boldsymbol{A}/K](f', g'))$ where e is the identity of G. Since $[\boldsymbol{A}/H \times \boldsymbol{A}/K](f', g')$ is strongly connected, there exists $\beta \in G$ such that $\psi((H, Kr)) = (Hf'\beta, Kg'\beta)$. Since $\psi \in \mathrm{Hom}([\boldsymbol{A}/H \times \boldsymbol{A}/K](e, r), [\boldsymbol{A}/H \times \boldsymbol{A}/K](f', g'))$, $\psi((H, Kr)z^{A/H \times A/K}) = \psi((H, Kr))z^{A/H \times A/K}$. Consequently, we have $(Hf'\beta, Kg'\beta) = (Hf'\beta x, Kg'\beta x)$. That is, $Hf'\beta = Hf'\beta x$ and $Kg'\beta = Kg'\beta x$. Put $\alpha = f'\beta$ (since β depends on r, α depends on r). Then we have $x \in \alpha^{-1}(H \cap s^{-1}Ks)\alpha$. Thus we have $H \cap r^{-1}Kr \subseteq \alpha^{-1}(H \cap s^{-1}Ks)\alpha$.

Theorem 3.3.1 *For any X $(|X| \geq 2)$, $V_X(S)$ is not a lattice.*

Proof To prove the theorem, it is enough to find groups G, H and K which do not satisfy the condition in the above lemma. Let $G = S_5 =$

$\langle\{(1\,2),(2\,3\,4\,5)\}\rangle$, $H = \langle\{(1\,2),(3\,4\,5)\}\rangle$ and $K = \langle\{(1\,3),(3\,4\,5)\}\rangle$ where by $\langle E\rangle$ we denote the group generated by E.

We prove that $H \cap s^{-1}Ks \neq H$ for any $s \in G$. Let $\sigma \in K$. Then $\sigma(2) = 2$. Let $s \in G = S_5$. There exists $i = 1, 2, \ldots, 5$ such that $s(i) = 2$. hence $s^{-1}\sigma s(i) = s^{-1}\sigma(2) = s^{-1}(2) = i$. Consequently, $(1\,2)(3\,4\,5) \in H$ but $(1\,2)(3\,4\,5) \notin s^{-1}Ks$. Therefore, $|H \cap s^{-1}Ks| \leq 3$. If $H \cap s^{-1}Ks \neq \langle\{(3\,4\,5)\}\rangle$, then $|\alpha^{-1}(H \cap s^{-1}Ks)\alpha| \leq 2$ for any $\alpha \in G$. In this case, we can take e as r and $|H \cap r^{-1}Kr| = |H \cap K| = |\langle\{(3\,4\,5)\}\rangle| = 3$. Thus $H \cap r^{-1}Kr \subseteq \alpha^{-1}(H \cap s^{-1}Ks)\alpha$ does not hold for any $\alpha \in G$. On the other hand, if $H \cap s^{-1}Ks = \langle\{(3\,4\,5)\}\rangle$, then $H \cap s^{-1}Ks$ is a cyclic group of order 3 and hence $\alpha^{-1}(H \cap s^{-1}Ks)\alpha$ is also a cyclic group of order 3 for any $\alpha \in G$. In this case, we can take $(2\,3) \in G$ as r. Then $r^{-1}Kr \ni (2\,3)(1\,3)(2\,3) = (1\,2)$ and $H \cap r^{-1}Kr = \langle\{(1\,2)\}\rangle$. Since this group is a cyclic group of order 2, $H \cap r^{-1}Kr \subseteq \alpha^{-1}(H \cap s^{-1}Ks)\alpha$ does not hold for any $\alpha \in G$. Hence the above groups G, H and K do not satisfy the condition in the above lemma.

Corollary 3.3.1 *Let $|X| \geq 2$ and let $V_X(P)$ be a class such that $V_X(S) \subseteq V_X(P)$. Then $V_X(P)$ is not a lattice.*

Proof By Corollary 3.2.3 and Theorem 3.3.1, there exist $\boldsymbol{A}, \boldsymbol{B} \in V_X(S)$ such that $\left|[\boldsymbol{A} \circ \boldsymbol{B}]_{V_X(S)}\right| \geq 2$. Then we have $\boldsymbol{A}, \boldsymbol{B} \in V_X(P)$ and, by Proposition 3.2.2, we have $\left|[\boldsymbol{A} \circ \boldsymbol{B}]_{V_X(P)}\right| \geq \left|[\boldsymbol{A} \circ \boldsymbol{B}]_{V_X(S)}\right| \geq 2$. This means that $V_X(P)$ is not a lattice.

From the above result, $V_X(Au)$ and $V_X(C)$ are not lattices if $|X| \geq 2$. For the case that $|X| = 1$, it can be shown that $V_X(C)$ is a lattice, but $V_X(Au)$ is not so.

3.4 Classes which form lattices

In the previous section, we showed that some classes are not lattices. In this section, we will deal with three classes which are lattices. One class is not a strong class, i.e. the set of all quasiperfect automata.

The other two classes are strong classes, i.e. the set of all perfect automata and the set of all strongly directable automata.

The following result is obvious.

Lemma 3.4.1 *Let $V(P)$ be a class such that $[\boldsymbol{A} * \boldsymbol{B}]_{V(P)} \neq \emptyset$ for any $\boldsymbol{A}, \boldsymbol{B} \in V(P)$. If $V(P)$ is an upper semilattice, i.e. $[\boldsymbol{A} \circ \boldsymbol{B}]_{V(P)}$ is a singleton set for all $\boldsymbol{A}, \boldsymbol{B} \in V(P)$, then $V(P)$ is necessarily a lattice.*

Remark 3.4.1 There is a class $V(P)$ such that $\mathbf{0} \in V(P)$ where $\mathbf{0}$ is the 1-state automaton. For instance, a strong class is one such. Notice that, for such a class, we have $[\boldsymbol{A} * \boldsymbol{B}]_{V(P)} \neq \emptyset$ for any $\boldsymbol{A}, \boldsymbol{B} \in V(P)$ and therefore, to prove that $V(P)$ forms a lattice, it is enough to show that $V(P)$ is an upper semilattice.

In what follows, we do not distinguish the element of a singleton set from a singleton set itself. Now we prepare to handle the class of all quasiperfect X-automata.

Lemma 3.4.2 *Let $\boldsymbol{A} = (S, X)$ and $\boldsymbol{B} = (T, X)$ be strongly connected automata. If one of the following three conditions is satisfied, then $[\boldsymbol{A} \circ \boldsymbol{B}]_{V(S)}$ is a singleton set: (i) $\boldsymbol{A}$ is a quasiperfect automaton. (ii) $\boldsymbol{B}$ is a quasiperfect automaton. (iii) $\boldsymbol{A} \times \boldsymbol{B}$ is a transitive automaton.*

Proof Since the idea is the same for any case, we prove only the case (i). First, it follows from Corollary 3.2.3 that $[\boldsymbol{A} \circ \boldsymbol{B}]_{V(S)} \neq \emptyset$. Assume $\boldsymbol{C}_1, \boldsymbol{C}_2 \in [\boldsymbol{A} \circ \boldsymbol{B}]_{V(S)}$. Here, notice that $\boldsymbol{C}_i = (U_i, X), i = 1, 2$, is a strongly connected subautomaton of $\boldsymbol{A} \times \boldsymbol{B}$. By the strong connectedness of $\boldsymbol{B}$, we can choose $s, s'' \in S$ and $t \in T$ such that $(s, t) \in U_1$ and $(s'', t) \in U_2$. Since $\boldsymbol{A}$ is transitive, there exists some $f \in \mathrm{Aut}(\boldsymbol{A})$ such that $s'' = f(s)$. We define the mapping ρ of U_1 into U_2 as $\rho((s', t')) = (f(s'), t')$ for any $(s', t') \in U_1$. Now, we prove that ρ is a homomorphism of $\boldsymbol{C}_1$ onto $\boldsymbol{C}_2$. By the strong connectedness of $\boldsymbol{C}_1$, for any $(s', t') \in U_1$ there exists some $u \in X^*$ such that $(s, t)u^{A \times B} = (s', t')$. Namely, $su^A = s'$ and $tu^B = t'$. By $f \in \mathrm{Aut}(\boldsymbol{A})$, we have $f(s') = f(su^A) = f(s)u^A = s''u^A$. Therefore, we have $(f(s'), t') = (s'', t)u^{A\times B} \in U_2$. Thus we have verified that ρ is a mapping of U_1 onto U_2. On the other hand, for any $(s', t') \in U_1$ and all $a \in X$, we have $\rho((s', t')a^{C_1}) = \rho((s', t')a^{A\times B}) = \rho((s'a^A, t'a^B)) = (f(s'a^A), t'a^B) = (f(s')a^A, t'a^B) = (f(s'), t')a^{A\times B} = \rho((s', t'))a^{C_2}$. This means that ρ is a homomorphism of $\boldsymbol{C}_1$ onto $\boldsymbol{C}_2$. That is, we have $\boldsymbol{C}_1 \geq \boldsymbol{C}_2$. By $\boldsymbol{C}_1, \boldsymbol{C}_2 \in [\boldsymbol{A} \circ \boldsymbol{B}]_{V(S)}$, this implies $\boldsymbol{C}_1 = \boldsymbol{C}_2$. Consequently, $[\boldsymbol{A} \circ \boldsymbol{B}]_{V(S)}$ is a singleton set.

We denote the set of all quasiperfect automata by $V_X(Qp)$. As will be shown in Theorem 3.4.1, $V_X(Qp)$ is a lattice but it is not a strong class.

Proposition 3.4.1 *For any X ($|X| \geq 2$), $V_X(Qp)$ is not a strong class.*

Proof Notice that there are a finite group G and its subgroup H such that H is not a normal subgroup of G. Consider a group-type automaton $\boldsymbol{A} = (G, X)$ and automaton $\boldsymbol{A}/H$. Then, by Theorem 3.1.1, $\boldsymbol{A} \in V(Qp)$ and $\boldsymbol{A}/H \leq \boldsymbol{A}$. Suppose that $V(Qp)$ is a strong class. Then $\boldsymbol{A}/H = (G/H, X) \in V(Qp)$. Since $\boldsymbol{A}$ is a group-type automaton, for any $h \in H$ there exists $u \in X^*$ such that $(Hg)u^{A/H} = (Hg)h$ for any $g \in G$. Let $g \in G$. Since $\boldsymbol{A}/H$ is transitive, there exists $\phi \in Aut(\boldsymbol{A}/H)$ such that $\phi(H) = Hg$. Consider $\phi(Hu^{A/H})$. Then $\phi(Hu^{A/H}) = \phi(H)u^{A/H} = Hgh$. On the other hand, $\phi(Hu^{A/H}) = \phi(H) = Hg$, i.e. $Hg = Hgh$. This implies that $H = g^{-1}Hg$. Consequently, H is a normal subgroup of G, which is a contradiction. Therefore, $V(Qp)$ is not a strong class.

In general, if $V(P)$ is not a strong class, then $[\boldsymbol{A} \circ \boldsymbol{B}]_{V(P)}$ does not consist of subautomata of $\boldsymbol{A} \times \boldsymbol{B}$. Hence, in this case, we have some difficulty in computing $[\boldsymbol{A} \circ \boldsymbol{B}]_{V(P)}$. However, for the class $V(Qp)$, we can show that it is a lattice.

Lemma 3.4.3 *For any $\boldsymbol{A}, \boldsymbol{B} \in V(Qp)$, $[\boldsymbol{A} \circ \boldsymbol{B}]_{V(S)}$ is an element of $V(Qp)$. Notice that, by Lemma 3.4.2, $[\boldsymbol{A} \circ \boldsymbol{B}]_{V(S)}$ is a singleton set and therefore it is considered as an element.*

Proof By Lemma 3.4.2, $[\boldsymbol{A} \circ \boldsymbol{B}]_{V(S)}$ is a singleton set for any $\boldsymbol{A} = (S, X), \boldsymbol{B} = (T, X)$. Let $(f, g) \in \text{Aut}(\boldsymbol{A}) \times \text{Aut}(\boldsymbol{B})$. We define the mapping (f, g) as $(f, g)((s, t)) = (f(s), g(t))$ for any $(s, t) \in S \times T$. Then we have $(f, g) \in \mathrm{Aut}(\boldsymbol{A} \times \boldsymbol{B})$. Thus, we have $\mathrm{Aut}(\boldsymbol{A}) \times \mathrm{Aut}(\boldsymbol{B}) \subseteq \mathrm{Aut}(\boldsymbol{A} \times \boldsymbol{B})$. This implies that $\boldsymbol{A} \times \boldsymbol{B}$ is a transitive automaton. Let $\boldsymbol{C} = [\boldsymbol{A} \circ \boldsymbol{B}]_{V(S)}$. Since $\boldsymbol{A} \times \boldsymbol{B}$ is transitive and $\boldsymbol{C}$ is a strongly connected subautomaton of $\boldsymbol{A} \times \boldsymbol{B}$, $\boldsymbol{C}$ is transitive, i.e. $\boldsymbol{C} = [\boldsymbol{A} \circ \boldsymbol{B}]_{V(S)} \in V(Qp)$.

Theorem 3.4.1 *$V(Qp)$ is a lattice.*

Proof First notice that $\mathbf{0} \in V(Qp)$. By Remark 3.4.1, to prove the theorem, it is enough to show that, for any $\boldsymbol{A}, \boldsymbol{B} \in V(Qp)$, $[\boldsymbol{A} \circ \boldsymbol{B}]_{V(Qp)}$ is a singleton set. To this end, we show first that $[\boldsymbol{A} \circ \boldsymbol{B}]_{V(Qp)} \subseteq [\boldsymbol{A} \circ \boldsymbol{B}]_{V(S)}$. Let $\boldsymbol{C} \in [\boldsymbol{A} \circ \boldsymbol{B}]_{V(Qp)}$. Then we have $\boldsymbol{C} \geq \boldsymbol{A}, \boldsymbol{B}$ and $\boldsymbol{C} \in V(S)$. Consequently, there exists some $\boldsymbol{D} \in [\boldsymbol{A} \circ \boldsymbol{B}]_{V(S)}$ such that $\boldsymbol{C} \geq \boldsymbol{D} \geq \boldsymbol{A}, \boldsymbol{B}$. By Lemma 3.4.3, we have $\boldsymbol{D} \in V(Qp)$ and thus we have $\boldsymbol{C} = \boldsymbol{D}$. Namely, we have $\boldsymbol{C} \in [\boldsymbol{A} \circ \boldsymbol{B}]_{V(S)}$. From this, we have $[\boldsymbol{A} \circ \boldsymbol{B}]_{V(Qp)} \subseteq [\boldsymbol{A} \circ \boldsymbol{B}]_{V(S)}$. This means that $[\boldsymbol{A} \circ \boldsymbol{B}]_{V(Qp)}$ has at most one element. But, $[\boldsymbol{A} \circ \boldsymbol{B}]_{V(S)}$ is a singleton set and, by Lemma 3.4.3, it is an element of $V(Qp)$. That is, we have $[\boldsymbol{A} \circ \boldsymbol{B}]_{V(Qp)} \neq \emptyset$. Therefore, $[\boldsymbol{A} \circ \boldsymbol{B}]_{V(Qp)}$ is a singleton set.

Now, we deal with the class of perfect automata. By $V_X(Pe)$ we denote the set of all perfect X-automata. The following result is obvious.

Proposition 3.4.2 *$V_X(Pe)$ is a strong class.*

Theorem 3.4.2 *$V(Pe)$ is a lattice.*

Proof By Theorem 3.1.2, we have $V(Pe) \subseteq V(Qp)$. Moreover, by the above proposition, $V(Pe)$ is a strong class. Therefore, by Proposition 3.2.2 and Theorem 3.4.1, to prove that $V(Pe)$ is a lattice, it is enough to show that for any $\boldsymbol{A}, \boldsymbol{B} \in V(Pe)$ we have $[\boldsymbol{A} \circ \boldsymbol{B}]_{V(Pe)} \neq \emptyset$. By the commutativities of $\boldsymbol{A}$ and $\boldsymbol{B}$, $\boldsymbol{A} \times \boldsymbol{B}$ is abelian. Therefore, its subautomaton $[\boldsymbol{A} \circ \boldsymbol{B}]_{V(S)}$ is also abelian. Consequently, we have $[\boldsymbol{A} \circ \boldsymbol{B}]_{V(S)} \in V(Pe)$. On the other hand, we have $[\boldsymbol{A} \circ \boldsymbol{B}]_{V(S)} \geq \boldsymbol{A}, \boldsymbol{B}$. Thus, we have $[\boldsymbol{A} \circ \boldsymbol{B}]_{V(Pe)} \neq \emptyset$.

Definition 3.4.1 An automaton $\boldsymbol{A} = (S, X)$ is said to be *directable* if the following condition is satisfied: For any $s, t \in S$, there exists some $u \in X^*$ such that $su^A = tu^A$.

By $V_X(Dir)$ we denote the set of all directable automata.

The following two results are obvious.

Proposition 3.4.3 *$V_X(Dir)$ is a strong class.*

Proposition 3.4.4 *For any $\boldsymbol{A}, \boldsymbol{B} \in V_X(Dir)$, we have $\boldsymbol{A} \times \boldsymbol{B} \in V_X(Dir)$.*

From the above result, we have:

Proposition 3.4.5 *For any $\boldsymbol{A}, \boldsymbol{B} \in V(Dir)$, we have $[\boldsymbol{A} \circ \boldsymbol{B}]_{V(Dir)} \neq \emptyset$.*

However, $V_X(Dir)$ is not a lattice. The reader can give an example that there are directable automata $\boldsymbol{A}$ and $\boldsymbol{B}$ with $\left|[\boldsymbol{A} \circ \boldsymbol{B}]_{V(Dir)}\right| \geq 2$ for any $X(|X| \geq 2)$. Now, consider a subclass of $V_X(Dir)$.

Definition 3.4.2 A directable automaton is said to be *strongly directable* if it is strongly connected. By $V_X(Sdir)$ we denote the set of all strongly directable X-automata.

The following result is obvious.

Proposition 3.4.6 *$V_X(Sdir)$ is a strong class.*

Theorem 3.4.3 *$V_X(Sdir)$ is a lattice.*

Proof Let $\boldsymbol{A} = (S, X)$ and $\boldsymbol{B} = (T, X)$ be elements of $V(Sdir)$. By Proposition 3.4.4, we have $\boldsymbol{A} \times \boldsymbol{B} \in V(Dir)$. Therefore, for any $\boldsymbol{C} \in [\boldsymbol{A} \circ \boldsymbol{B}]_{V(S)}$, we have $\boldsymbol{C} \in V(Dir) \cap V(S) = V(Sdir)$. From this, we have $[\boldsymbol{A} \circ \boldsymbol{B}]_{V(SDir)} \neq \emptyset$. On the other hand, since $V(Sdir)$ is a strong class, to prove that $[\boldsymbol{A} \circ \boldsymbol{B}]_{V(Sdir)}$ is a singleton, by Proposition 3.2.2, we need only to show that $[\boldsymbol{A} \circ \boldsymbol{B}]_{V(S)}$ is a singleton. Let $\boldsymbol{C}_1, \boldsymbol{C}_2 \in [\boldsymbol{A} \circ \boldsymbol{B}]_{V(S)}$, let $\boldsymbol{C}_1 = (U_1, X)$ and let $\boldsymbol{C}_2 =$

(U_2, X) where $\boldsymbol{C}_1$ and $\boldsymbol{C}_2$ are strongly connected subautomata of $\boldsymbol{A} \times \boldsymbol{B}$. For any $(s,t) \in U_1$ and all $(s',t') \in U_2$, by the fact that $\boldsymbol{A} \times \boldsymbol{B}$ is directable, there exists some $u \in X^*$ such that $(s,t)u^{C_1} = (s,t)u^{A \times B} = (s',t')u^{A \times B} = (s',t')u^{C_2}$ $(= (s'',t''))$. Notice that each element of U_1 (of U_2) does not go out from U_1 (from U_2) by any input. This is because $\boldsymbol{C}_1$ and $\boldsymbol{C}_2$ are strongly connected subautomata of $\boldsymbol{A} \times \boldsymbol{B}$. Thus we have $(s'',t'') \in U_1 \cap U_2$. Consequently, we have $U_1 \cap U_2 \neq \emptyset$. This means that we have $\boldsymbol{C}_1 = \boldsymbol{C}_2$.

To conclude this section, we provide the following algorithm.

Algorithm Let $V_X(P)$ be one of $V_X(VQp)$, $V_X(Pe)$ and $V_X(Sdir)$. Moreover, let $\boldsymbol{A}, \boldsymbol{B} \in V_X(P)$. Then we can obtain the least upper bound of $\{\boldsymbol{A}, \boldsymbol{B}\}$ in $V_X(P)$ as follows: (i) Construct $\boldsymbol{A} \times \boldsymbol{B}$. (ii) Let $S[X]$ be the set of all strongly connected X-subautomata of $\boldsymbol{A} \times \boldsymbol{B}$. (iii) Let $\boldsymbol{C}$ be an automaton (from $S[X]$) whose cardinality of state set is minimum. (iv) Then $\boldsymbol{C}$ is the least upper bound of $\{\boldsymbol{A}, \boldsymbol{B}\}$ in $V_X(P)$.

3.5 Computation of $[\boldsymbol{A} * \boldsymbol{B}]_{V(P)}$

In the previous section, some examples of classes which are lattices were given. In the present section, we study how to compute the value $[\boldsymbol{A} * \boldsymbol{B}]_{V(P)}$, $\boldsymbol{A}, \boldsymbol{B} \in V(P)$ for a class $V(P)$ which is a lattice.

First, let $\boldsymbol{C} = (U, X)$ be a subautomaton of $\boldsymbol{A} \times \boldsymbol{B} = (S \times T, X)$, denoted by $\boldsymbol{C} \subset \boldsymbol{A} \times \boldsymbol{B}$, such that $\pi_A(U) = S$ and $\pi_B(U) = T$. For $(s,t), (s',t') \in U$, $(s,t) \sim (s',t')$ means that we have $s = s'$ or $t = t'$. And $(s,t) \approx (s',t')$ means that there exists some $n \geq 2$ such that $(s,t) = (s_1,t_1) \sim (s_2,t_2) \sim \cdots \sim (s_{n-1},t_{n-1}) \sim (s_n,t_n) = (s',t')$. Moreover, for $s, s' \in S$ $(t, t' \in T)$, $s \leftrightarrow s'$ $(t \leftrightarrow t')$ means that there exist some $t, t' \in T$ $(s, s' \in S)$ such that $(s,t) \approx (s',t')$.

The following result is obvious.

Lemma 3.5.1 *The relation $\leftrightarrow$ is a congruence relation on S (on T). That is, for any $u \in X^*$ we have $su^A \leftrightarrow s'u^A$ $(tu^B \leftrightarrow t'u^B)$ if $s \leftrightarrow s'$ $(t \leftrightarrow t')$.*

Now, for any $s \in S$ and $t \in T$, we put $\widehat{s} = \{s' \mid s \leftrightarrow s'\}$ and $\widehat{t} = \{t' \mid t \leftrightarrow t'\}$. Then we have the automaton $\widehat{\boldsymbol{A}} = (\{\widehat{s} \mid s \in S\}, X)$ and the automaton $\widehat{\boldsymbol{B}} = (\{\widehat{t} \mid t \in T\}, X)$ such that $\widehat{s}a^{\widehat{A}} = \widehat{sa^A}$ for any $(s, a) \in S \times X$ and $\widehat{t}a^{\widehat{B}} = \widehat{ta^B}$ for any $(t, x) \in T \times X$. By Lemma 3.5.1, $\widehat{\boldsymbol{A}}$ and $\widehat{\boldsymbol{B}}$ are automata such that $\boldsymbol{A} \geq \widehat{\boldsymbol{A}}$ and $\boldsymbol{B} \geq \widehat{\boldsymbol{B}}$.

Lemma 3.5.2 $\widehat{\boldsymbol{A}} = \widehat{\boldsymbol{B}}$.

Proof For any $s \in S$, we put $\rho(\widehat{s}) = \widehat{t}$ if there exists some $(s, t) \in U$. We show that ρ is well defined as a mapping. Let $\widehat{s} = \widehat{s'}$ where $(s, t), (s', t') \in U$. Now, we prove that $\widehat{t} = \widehat{t'}$. By $\widehat{s} = \widehat{s'}$, there exist some $(s, r), (s', r') \in U$ such that $(s, r) \approx (s', r')$. Notice that we have $(s, t) \sim (s, r)$ and $(s', r') \sim (s', t')$. Therefore, we have $(s, t) \approx (s', t')$. That is, we have $\widehat{t} = \widehat{t'}$. Next, by $\pi_B(U) = T$, ρ is a surjection. Now, we prove that ρ is a homomorphism. First, we put $s \in S$, $a \in X$ and $\rho(\widehat{s}) = \widehat{t}$. Note that $\rho(\widehat{s}) = \widehat{t}$ implies $(s, t) \in U$ and thus $(sa^A, ta^B) \in U$. The last one implies $\rho(\widehat{sa^A}) = \widehat{ta^B}$. Consequently, we have the following equality: $\rho(\widehat{s}a^{\widehat{A}}) = \rho(\widehat{sa^A}) = \widehat{ta^B} = \widehat{t}a^{\widehat{B}} = \rho(\widehat{s})a^{\widehat{B}}$. This means $\widehat{\boldsymbol{A}} \geq \widehat{\boldsymbol{B}}$. In the same way, we have $\widehat{\boldsymbol{B}} \geq \widehat{\boldsymbol{A}}$. Therefore, we have $\widehat{\boldsymbol{A}} = \widehat{\boldsymbol{B}}$.

Definition 3.5.1 By $\delta(\boldsymbol{C})$, we denote $\widehat{\boldsymbol{A}}$ $(= \widehat{\boldsymbol{B}})$.

Thus we have the following:

Theorem 3.5.1 *$\boldsymbol{A} \geq \delta(\boldsymbol{C})$ and $\boldsymbol{B} \geq \delta(\boldsymbol{C})$.*

Theorem 3.5.2 *Let $\boldsymbol{D} = (W, X)$ be an automaton such that $\boldsymbol{A} \geq \boldsymbol{D}$ and $\boldsymbol{B} \geq \boldsymbol{D}$. Then there exists a subautomaton $\boldsymbol{C}$ of $\boldsymbol{A} \times \boldsymbol{B}$ such that $\boldsymbol{D} = \delta(\boldsymbol{C})$.*

Before proving the above theorem, we provide the definitions of a pullback system and a pullback automaton.

Definition 3.5.2 Let $\boldsymbol{D}$ be an automaton such that $\boldsymbol{A} \geq \boldsymbol{D}$ and $\boldsymbol{B} \geq \boldsymbol{D}$, and f and g be elements of $\mathrm{Hom}(\boldsymbol{A}, \boldsymbol{D})$ and of $\mathrm{Hom}(\boldsymbol{B}, \boldsymbol{D})$, respectively. Then a 5-tuple $X = (\boldsymbol{A}, \boldsymbol{B}, f, g, \boldsymbol{D})$ is called a *pullback system* and the automaton $\wp[X] = (V, X)$ induced by it is called the *pullback automaton* of X. Here $V = \{(s,t) \in S \times T \mid f(s) = g(t)\}$ and the state transition function $\delta_{\wp[X]}$ is defined as $(s,t)a^{\wp[X]} = (sa^A, ta^B)$ for any $(s,t) \in V$ and $a \in X$. Then $\wp[X]$ is a subautomaton of $\boldsymbol{A} \times \boldsymbol{B}$ such that $\pi_A(V) = S$ and $\pi_B(V) = T$.

Proof of Theorem 3.5.2 Let $X = (\boldsymbol{A}, \boldsymbol{B}, f, g, \boldsymbol{D})$ be a pullback system and let $\wp[X]$ be its pullback automaton. Put $\boldsymbol{C} = \wp[X]$. Now, we prove that $\delta(\boldsymbol{C}) = \boldsymbol{D}$. Let apply the δ-operator to $\boldsymbol{C} = \wp[X]$. Notice that for any $s, s' \in S$ the following three conditions are equivalent: (i) $\widehat{s} = \widehat{s'}$. (ii) There exist some $t, t' \in T$ such that $(s,t) \approx (s',t')$. (iii) There exist some $t, t' \in T$ such that $f(s) = g(t) = f(s') = g(t')$.

From the above, it follows that $\rho(\widehat{s}) = f(s), s \in S$ is well defined as a bijection of the state set of $\widehat{\boldsymbol{A}}$ onto W. By $f \in \mathrm{Hom}(\boldsymbol{A}, \boldsymbol{D})$, it can be proved that ρ is a homomorphism as follows: $\rho(\widehat{s}a^{\widehat{A}}) = \rho(\widehat{sa^A}) = f(sa^A) = f(s)a^D = \rho(\widehat{s})a^D$ for any $(s,a) \in S \times X$.

We will use the above result for obtaining $[\boldsymbol{A} * \boldsymbol{B}]_{V(P)}$.

Lemma 3.5.3 *Let $\boldsymbol{C} = (U, X)$ and $\boldsymbol{C}' = (U', X)$ be automata such that $\pi_A(U') = S$, $\pi_B(U') = T$ and $\boldsymbol{C}' \subset \boldsymbol{C} \subset \boldsymbol{A} \times \boldsymbol{B}$. Then we have $\delta(\boldsymbol{C}') \geq \delta(\boldsymbol{C})$.*

Proof For any $s \in S$, by $\widehat{s}$ we denote the equivalence class in U containing s by $\leftrightarrow$, and by $\widehat{\widehat{s}}$ we denote the equivalence class in U' containing s by $\leftrightarrow$. For $s, s' \in S$, we assume that $\widehat{\widehat{s}} = \widehat{\widehat{s'}}$. That is, there exist some $t, t' \in T$ such that $(s,t), (s',t') \in U'$ and $(s,t) \approx (s',t')$. By $\boldsymbol{C}' \subset \boldsymbol{C}$, we have $(s,t), (s',t') \in U$ and $(s,t) \approx (s',t')$. Thus, we have $\widehat{s} = \widehat{s'}$. Consequently, $\rho(\widehat{\widehat{s}}) = \widehat{s}$ $(s \in S)$ is well defined as a mapping. It is obvious that this mapping is a surjection. Since $\widehat{\widehat{s}}$ and $\widehat{s}$ are equivalence classes by congruence relations, respectively and $\widehat{\widehat{s}} \subseteq \widehat{s}$ $(s \in S)$, ρ is a homomorphism.

From the above result, for any strong class $V(P)$, we have the following:

Proposition 3.5.1 $[\boldsymbol{A} * \boldsymbol{B}]_{V(P)} \subseteq \{\delta(\boldsymbol{C}) \mid \boldsymbol{C} = (U, X) \subset \boldsymbol{A} \times \boldsymbol{B}, \pi_A(U) = S,\ \pi_B(U) = T$ *and* $\boldsymbol{C}' = \boldsymbol{C}$ *if* $\boldsymbol{C} \supset \boldsymbol{C}' = (U', X), \pi_A(U') = S$ *and* $\pi_B(U') = T\}$.

If $V(P) \subseteq V(S)$ in the above, then we can choose only strongly connected subautomata of $\boldsymbol{A} \times \boldsymbol{B}$ as $\boldsymbol{C}$. Moreover, if $V(P)$ forms a lattice and two arbitrary strongly connected subautomata of $\boldsymbol{A} \times \boldsymbol{B}$ are isomorphic to each other ($= [\boldsymbol{A} \circ \boldsymbol{B}]_{V(P)}$) through an element of $\mathrm{Aut}(\boldsymbol{A}) \times \mathrm{Aut}(\boldsymbol{B})$, we can choose only $[\boldsymbol{A} \circ \boldsymbol{B}]_{V(P)}$ as $\boldsymbol{C}$. Thus we have the following result. Notice that the class $V(Qp)$ can be treated like other strong classes though it is not a strong class.

Theorem 3.5.3 *Let $V(P)$ be one of $V(Qp)$, $V(Pe)$ and $V(Sdir)$. Then for any $\boldsymbol{A}, \boldsymbol{B} \in V(P)$ we have $[\boldsymbol{A} * \boldsymbol{B}]_{V(P)} = \delta([\boldsymbol{A} \circ \boldsymbol{B}]_{V(P)})$.*

Example 3.5.1 For the automata $\boldsymbol{A}, \boldsymbol{B} \in V(Sdir)$ in Fig. 3.1, compute $[\boldsymbol{A} \circ \boldsymbol{B}]_{V(Sdir)}$ and $[\boldsymbol{A} * \boldsymbol{B}]_{V(Sdir)}$.

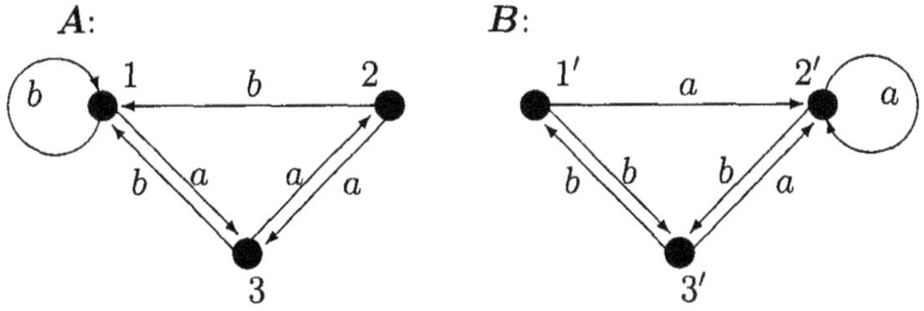

Figure 3.1: $\boldsymbol{A}$, $\boldsymbol{B}$

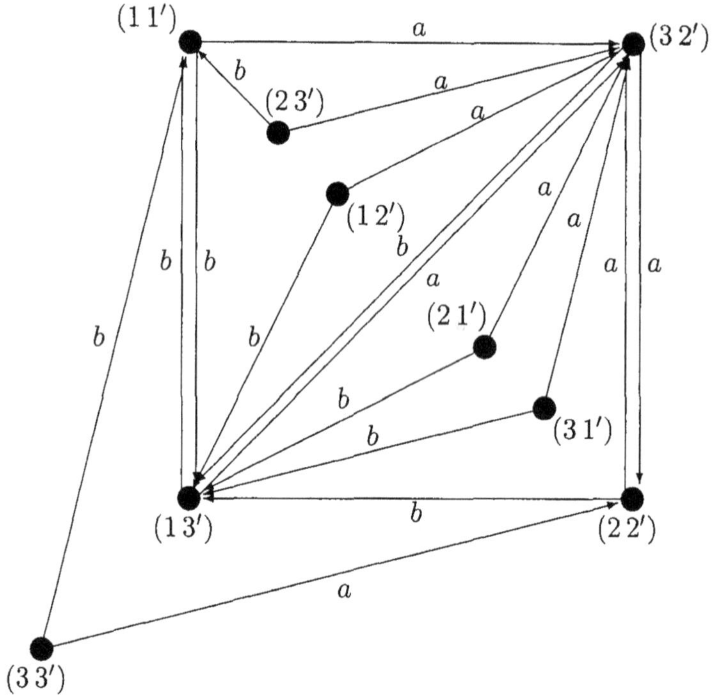

Figure 3.2: $\boldsymbol{A} \times \boldsymbol{B}$

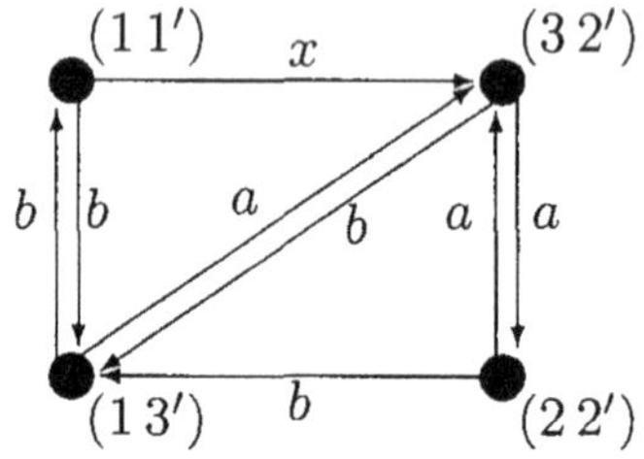

Figure 3.3: $[\boldsymbol{A} \circ \boldsymbol{B}]_{V(Sdir)}$

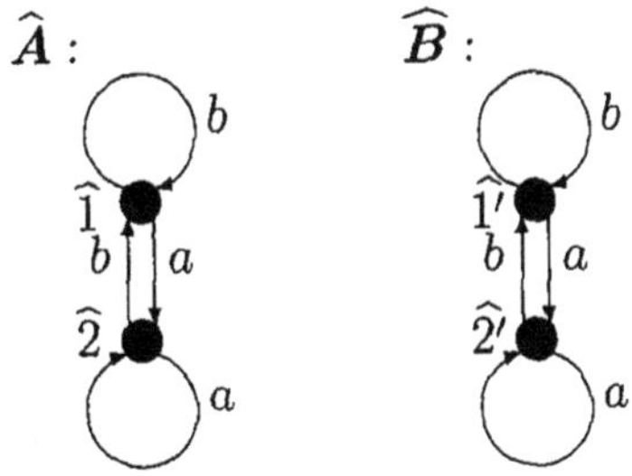

$(1,1') \approx (1,3') \quad (3,2') \approx (2,2')$

$\widehat{1} = \{1\} \qquad \widehat{1'} = \{1',3'\}$

$\widehat{2} = \{2,3\} \qquad \widehat{2'} = \{2'\}$

Figure 3.4: $\widehat{\boldsymbol{A}} = \widehat{\boldsymbol{B}} = \delta([\boldsymbol{A} \circ \boldsymbol{B}]_{V(Sdir)})$

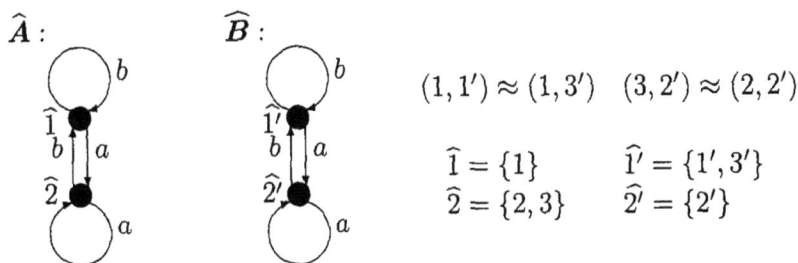

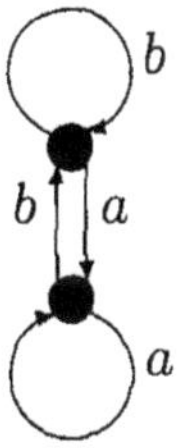

Figure 3.5: $[\boldsymbol{A} * \boldsymbol{B}]_{V(Sdir)}$

Chapter 4

Languages and Operations

A subset of the free monoid generated by a finite nonempty set is called a *language*. Our objective in this chapter is to study some languages generated (accepted) by grammars (acceptors). In Section 4.1, we will introduce regular and context-free grammars (finite and pushdown acceptors). These mechanisms define languages called *regular* and *context-free* languages. In Sections 4.2 and 4.3, we will consider operations on languages called *n-insertion* and *shuffle*. After providing a few properties of these operations, in Sections 4.4 and 4.5 we will deal with decompositions of regular languages into nontrivial regular languages by these operations. The existence of an algorithm to decide whether or not the shuffle closure of a regular language is regular is unknown. In Section 4.6, we will give a partial solution to this open problem.

The contents of this chapter are based on the results in [45], [38] and [27].

4.1 Grammars and acceptors

Let X^* denote the free monoid generated by a finite nonempty alphabet X and let $X^+ = X^* \setminus \{\epsilon\}$ where ϵ denotes the identity of X^*. For the sake of simplicity, if $X = \{a\}$, then we write a^+ and a^* instead of $\{a\}^+$ and $\{a\}^*$, respectively. Let $A \subseteq X^*$. Then A is called a *language* over X. By $|A|$, we denote the cardinality of A. If

$A \subseteq X^*$, then A^+ denotes the set of all concatenations of words in A and $A^* = A^+ \cup \{\epsilon\}$. In particular, if $A = \{w\}$, then we write w^+ and w^* instead of $\{w\}^+$ and $\{w\}^*$, respectively. Let $u \in X^*$. Then u is called a *word* over X. Especially, ϵ is called the *empty* word. If $u \in X^*$, then $|u|$ denotes the length of u. Moreover, if $a \in X$, then the number of the occurences of the letter a in the word u is denoted by $|u|_a$. Let $A \subseteq X^*$ be a language. Then the *principal congruence* P_A of A is defined as follows: For any $u, v \in X^*, u \equiv v(P_A)$ if and only if $(xuy \in A \Leftrightarrow xvy \in A)$ for any $x, y \in X^*$.

Definition 4.1.1 A *grammar* $\mathcal{G} = (V, X, P, S)$ consists of the following data: (1) V and X are finite nonempty sets, called the set of *variables* and an *alphabet*, respectively such that $V \cap X = \emptyset$. (2) P is a finite subset of $(X^*V^+X^*)^* \times (X \cup V)^*$, called the set of *production rules*. (3) $S \in V$ is called a *start symbol*. Moreover, a grammar $\mathcal{G} = (V, X, P, S)$ is said to be *regular* (*context-free*) if $P \subseteq (V \times (X \cup \{\epsilon\})V) \cup (V \times (X \cup \{\epsilon\})$ $(P \subseteq V \times (X \cup V)^*)$.

Let $\mathcal{G} = (V, X, P, S)$ be a grammar and let $(u, v) \in P$. Then usually we denote $u \to v$ instead of (u, v). Let $u \to v \in P$. Then $\alpha u \beta \Rightarrow \alpha v \beta$ for any $\alpha, \beta \in (X \cup V)^*$. By $\Rightarrow^*$, we denote the transitive closure of $\Rightarrow$, i.e. (1) $u \Rightarrow^* u$ for any $u \in (V \cup X)^*$, (2) $u \Rightarrow^* v$ if $u = u_0, u_1, u_2, \dots, u_{n-1}, u_n = v$ for some positive integer n and $u_i \Rightarrow u_{i+1}$ for any $i = 0, 1, \dots, n-1$. Then $u \Rightarrow^* v, u, v \in (V \cup X)^*$, is called a *derivation* from u to v. For a grammar $\mathcal{G} = (V, X, P, S)$, the *language* $\mathcal{L}(\mathcal{G})$ *generated by* G is defined as follows: $\mathcal{L}(\mathcal{G}) = \{u \in X^* \mid S \Rightarrow^* u\}$.

A language $L \subseteq X^*$ is said to be *regular* (*context-free*) if there exists a regular (context-free) grammar $\mathcal{G}$ such that $L = \mathcal{L}(\mathcal{G})$.

Remark 4.1.1 Consider a grammar $\mathcal{G} = (V, X, P, S)$ with $P \subseteq (V \times X^*V) \cup (V \times X^*)$. Then notice that $\mathcal{L}(\mathcal{G})$ is regular.

Remark 4.1.2 A context-free grammar $G = (V, X, P, S)$ is said to be in the *Greibach normal form* if $P \subseteq (V \times XV^*) \cup \{S \to \epsilon\}$. Notice that there exists a context-free grammar $\mathcal{G}$ in the Greibach normal form with $L = \mathcal{L}(\mathcal{G})$ for any context-free language $L \subseteq X^*$.

Now we define finite acceptors.

Definition 4.1.2 A *finite acceptor* $\boldsymbol{A} = (S, X, \delta, s_0, F)$ consists of the following data: (1) S is a finite nonempty set, called a *state set.* (2) X is a finite alphabet. (3) δ is a function, called a *state transition function,* of $S \times X$ into S. (4) $s_0 \in S$ is called the *initial state.* (5) $F \subseteq S$ is called the *set of final states.*

The state transition function δ can be extended to the function of $S \times X^*$ into S as follows: (1) $\delta(s, \epsilon) = s$ for any $s \in S$, (2) $\delta(s, au) = \delta(\delta(s, a), u)$ for any $s \in S, a \in X$ and $u \in X^*$.

Let $\boldsymbol{A} = (S, X, \delta, s_0, F)$ be a finite acceptor. Then the language $\mathcal{T}(\boldsymbol{A}) = \{u \in X^* \mid \delta(s_0, u) \in F\}$ is called the *language accepted* (or *recognized*) *by* $\boldsymbol{A}$.

As a more general case, we consider nondeterministic acceptors.

Definition 4.1.3 A *nondeterministic acceptor* $\boldsymbol{A} = (S, X, \delta, S_0, F)$ consists of the following data: (1) S, X are the same materials as in the definition of finite acceptors. (2) δ is a relation such that $\delta(s, a) \subseteq S$ for any $s \in S$ and $a \in X$. (3) $S_0 \subseteq S$ is called the *set of initial states.* (4) $F \subseteq S$ is called the *set of final states.*

As in the case of finite acceptors, δ can be extended to the following relation in a natural way, i.e. (1) $\delta(s, \epsilon) = \{s\}$ for any $s \in S$, (2) $\delta(s, au) = \bigcup_{t \in \delta(s,a)} \delta(t, u)$ for any $s \in S, a \in X$ and $u \in X^*$. Then the *language* $\mathcal{T}(\boldsymbol{A})$ *accepted* (or *recognized*) *by* $\boldsymbol{A}$ is defined as $\mathcal{T}(\boldsymbol{A}) = \{u \in X^* \mid \exists s_0 \in S_0, \delta(s_0, u) \cap F \neq \emptyset\}$.

Let $T \subseteq S$. We denote $\delta(T, u) = \bigcup_{t \in T} \delta(t, u)$ for $u \in X^*$. Thus $\mathcal{T}(\boldsymbol{A}) = \{u \in X^* \mid \delta(S_0, u) \cap F \neq \emptyset\}$.

Remark 4.1.3 In Definition 4.1.3, the state transition relation can be generalized as follows: δ is a relation such that $\delta(s, a) \subseteq S$ for any $s \in S$ and $a \in X \cup \{\epsilon\}$. Under this condition, a nondeterministic acceptor with $\delta(s, \epsilon) \setminus \{s\} \neq \emptyset$ for some $s \in S$ is called a *nondeterministic acceptor with* ϵ*-move.* Notice that, for a nondeterministic acceptor $\boldsymbol{A}$ with ϵ-move, there exists a nondeterministic acceptor $\boldsymbol{B}$ without ϵ-move such that $\mathcal{T}(\boldsymbol{A}) = \mathcal{T}(\boldsymbol{B})$.

Proposition 4.1.1 *Let $\boldsymbol{A}$ be a nondeterministic acceptor. Then there exists a finite acceptor $\overline{\boldsymbol{A}}$ such that $\mathcal{T}(\boldsymbol{A}) = \mathcal{T}(\overline{\boldsymbol{A}})$.*

Proof Let $\boldsymbol{A} = (S, X, \delta, S_0, F)$ be a nondeterministic acceptor. We construct the following finite acceptor $\overline{\boldsymbol{A}} = (\overline{S}, X, \overline{\delta}, s_0, \overline{F})$: (1) $\overline{S} = \{T \mid T \subseteq S\}$, (2) $s_0 = S_0$. (3) $\overline{\delta}(T, a) = \delta(T, a)$ for any $T \subseteq S$ and $a \in X$. (4) $\overline{F} = \{T \mid T \subseteq S, T \cap F \neq \emptyset\}$.

Now we prove that $\mathcal{T}(\boldsymbol{A}) = \mathcal{T}(\overline{\boldsymbol{A}})$. Let $u \in \mathcal{T}(\boldsymbol{A})$. Then $\delta(S_0, u) \cap F \neq \emptyset$. Hence $\overline{\delta}(s_0, u) \in \overline{F}$ and $u \in \mathcal{T}(\overline{\boldsymbol{A}})$. Consequently, $\mathcal{T}(\boldsymbol{A}) \subseteq \mathcal{T}(\overline{\boldsymbol{A}})$.

Conversely, let $u \in \mathcal{T}(\overline{A})$. Then $\overline{\delta}(s_0, u) \cap F \neq \emptyset$, i.e. $\delta(S_0, u) \cap F \neq \emptyset$. Hence $u \in \mathcal{T}(\boldsymbol{A})$ and $\mathcal{T}(\overline{\boldsymbol{A}}) \subseteq \mathcal{T}(\boldsymbol{A})$. Therefore, $\mathcal{T}(\boldsymbol{A}) = \mathcal{T}(\overline{\boldsymbol{A}})$. This completes the proof of the proposition.

Proposition 4.1.2 *Let $L \subseteq X^*$ be a language over X. Then L is regular if and only if L is accepted by a finite acceptor.*

Proof ($\Rightarrow$) Let $L \subseteq X^*$ be a regular language and let $\mathcal{G} = (V, X, P, S)$ be a regular grammar with $L = \mathcal{L}(\mathcal{G})$. We construct the following nondeterministic acceptor $\boldsymbol{A} = (T, X, \delta, \{t_0\}, F)$: (1) $T = V \cup \{\#\}$ with $\# \notin V$, (2) $B \in \delta(A, a)$ if $(A \to aB) \in P$ for any $a \in X$, and $A, B \in V$, (3) $\# \in \delta(C, a)$ if $(C \to a) \in P$ for any $a \in X$ and $C \in V$, (4) $t_0 = S$, (5) $F = \{\#\} \cup \{S\}$ if $(S \to \epsilon) \in P$ and $F = \{\#\}$ if $(S \to \epsilon) \notin P$.

First, we prove that $\mathcal{L}(\mathcal{G}) \subseteq \mathcal{T}(\boldsymbol{A})$. Let $u \in \mathcal{L}(\mathcal{G})$. If $u = \epsilon$, then $(S \to \epsilon) \in P$ and $t_0 = S \in F$. Hence $u = \epsilon \in \mathcal{T}(\boldsymbol{A})$. Assume $u \neq \epsilon$. Then $u = a_1a_2\cdots a_k, a_1, a_2, \ldots, a_k \in X$. Since $u \in \mathcal{L}(\mathcal{G})$, we have $\{S \to a_1T_1, T_1 \to a_2T_2, \ldots, T_{k-2} \to a_{k-1}T_{k-1}, T_{k-1} \to a_k\} \subseteq P$. Thus $\delta(t_0, a_1a_2\cdots a_k) \supseteq \delta(\delta(t_0, a_1a_2\cdots a_{k-1}), a_k) \supseteq \delta(T_{k-1}, a_k) \supseteq \{\#\}$, i.e. $\delta(t_0, a_1a_2\cdots a_k) \cap F \neq \emptyset$. This means that $u = a_1a_2\cdots a_k \in \mathcal{T}(\boldsymbol{A})$, i.e. $\mathcal{L}(\mathcal{G}) \subseteq \mathcal{T}(\boldsymbol{A})$.

Now we prove that $\mathcal{T}(\boldsymbol{A}) \subseteq \mathcal{L}(\mathcal{G})$. Let $u \in \mathcal{T}(\boldsymbol{A})$. If $u = \epsilon$, then $S \in F$ and $(S \to \epsilon) \in P$. Hence $u = \epsilon \in \mathcal{L}(\mathcal{G})$. Assume $u \neq \epsilon$. Then $u = a_1a_2\cdots a_k, a_1, a_2, \ldots, a_k \in X$. Since $u \in \mathcal{T}(\boldsymbol{A})$, there exist $T_1, T_2, \ldots, T_{k-1} \in V$ such that $T_1 \in \delta(S, a_1), T_2 \in \delta(T_1, a_2), \ldots, T_{k-1} \in \delta(T_{k-2}, a_{k-1}), \# \in \delta(T_{k-1}, a_k)$ or $S \in \delta(T_{k-1}, a_k)$. Thus $\{S \to a_1T_1, T_1 \to a_2T_2, \ldots, T_{k-2} \to a_{k-1}T_{k-1}, T_{k-1} \to a_k\} \subseteq P$ or $\{S \to a_1T_1, T_1 \to a_2T_2, \ldots, T_{k-2} \to a_{k-1}T_{k-1}, T_{k-1} \to$

$a_kS, S \to \epsilon\} \subseteq P$. Consequently, $S \Rightarrow^* a_1a_2\cdots a_k$ and hence $u = a_1a_2\cdots a_k \in \mathcal{L}(\mathcal{G})$, i.e. $\mathcal{T}(\boldsymbol{A}) \subseteq \mathcal{L}(\mathcal{G})$. Therefore, $\mathcal{T}(\boldsymbol{A}) = \mathcal{L}(\mathcal{G})$. By Proposition 4.1.1, there exists a finite acceptor $\overline{\boldsymbol{A}}$ such that $\mathcal{T}(\boldsymbol{A}) = \mathcal{T}(\overline{\boldsymbol{A}})$. Thus L is accepted by a finite acceptor.

($\Leftarrow$) Let $\boldsymbol{A} = (S, X, \delta, s_0, F)$ be a finite acceptor with $L = \mathcal{T}(\boldsymbol{A})$. We construct the following regular grammar: $\mathcal{G} = (V, X, P, s_0)$ where (1) $V = S$, (2) $P = P' \cup \{s_0 \to \epsilon\}$ if $s_0 \in F$ and $P = P'$ if $s_0 \notin F$ where $P' = \{s \to as' \mid a \in X, s, s' \in S, \delta(s,a) = s'\} \cup \{s \to a \mid a \in X, s \in S, \delta(s,a) \in F\}$.

We prove that $\mathcal{L}(\mathcal{G}) = \mathcal{T}(\boldsymbol{A})$. Let $u \in \mathcal{T}(\boldsymbol{A})$. If $u = \epsilon$, then $s_0 \in F$, i.e. $(s_0 \to \epsilon) \in P$. Hence $u \in \mathcal{L}(\mathcal{G})$. Assume $u \neq \epsilon$. Let $u = a_1a_2\cdots a_k, a_1, a_2, \ldots, a_k \in X$. Then there exist $s_0 = t_1, t_2, \ldots, t_k \in S$ such that $\delta(t_1, a_1) = t_2, \delta(t_2, a_2) = t_3, \ldots, \delta(t_{k-1}, a_{k-1}) = t_k, \delta(t_k, a_k) \in F$. This means that $\{s_0 \to a_1t_2, t_2 \to a_2t_3, \ldots, t_{k-1} \to a_{k-1}t_k, t_k \to a_k\} \subseteq P$. Hence $s_0 \Rightarrow^* a_1a_2\cdots a_k$ holds. Therefore, $u \in \mathcal{L}(\mathcal{G})$ and $\mathcal{T}(\boldsymbol{A}) \subseteq \mathcal{L}(\mathcal{G})$.

Now let $u \in \mathcal{L}(\mathcal{G})$. If $u = \epsilon$, then $(s_0 \to \epsilon) \in P$, i.e. $s_0 \in F$ and hence $u = \epsilon \in \mathcal{T}(\boldsymbol{A})$. Assume $u \neq \epsilon$. Then $u = a_1a_2\cdots a_k, a_1, a_2, \ldots, a_k \in X$ and $s_0 \Rightarrow^* a_1a_2\cdots a_k$. Thus $\{s_0 \to a_1t_2, t_2 \to a_2t_3, \ldots, t_{k-1} \to a_{k-1}t_k, t_k \to a_k\} \subseteq P$ or $\{s_0 \to a_1t_2, t_2 \to a_2t_3, \ldots, t_{k-1} \to a_{k-1}t_k, t_k \to a_ks_0, s_0 \to \epsilon\} \subseteq P$. Therefore, $\delta(s_0, a_1) = \delta(t_1, a_1) = t_2, \delta(t_2, a_2) = t_3, \ldots, \delta(t_{k-1}, a_{k-1}) = t_k, \delta(t_k, a_k) \in F$. Thus $\delta(s_0, a_1a_2\cdots a_k) \in F$, i.e. $u = a_1a_2\cdots a_k \in \mathcal{T}(\boldsymbol{A})$. Hence $\mathcal{L}(\mathcal{G}) \subseteq \mathcal{T}(\boldsymbol{A})$ and $\mathcal{L}(\mathcal{G}) = \mathcal{T}(\boldsymbol{A})$. Therefore, L is regular.

Proposition 4.1.3 *Let $L \subseteq X^*$ be a language over X. Then L is regular if and only if the principal congruence P_L is of finite index, i.e. the number of congruence classes of P_L is finite.*

Proof ($\Rightarrow$) Let $L \subseteq X^*$ be a regular language. By Proposition 4.1.2, there exists a finite acceptor $\boldsymbol{A} = (S, X, \delta, s_0, F)$ such that $L = \mathcal{T}(\boldsymbol{A})$. For any $u \in X^*$, we define the mapping f_u of S into S as $f_u(s) = \delta(s, u)$ For any $s \in S$. We show that $f_u = f_v$ implies $u \equiv v(P_L)$ for any $u, v \in X^*$. Let $xuy \in L$ for $x, y \in X^*$. Then $\delta(s_0, xvy) = \delta(\delta(\delta(s_0, x), v), y) = \delta(f_v(\delta(s_0, x)), y) = \delta(f_u(\delta(s_0, x)), y) = \delta(\delta(\delta(s_0, x), u), y) = \delta(s_0, xuy) \in F$. Therefore, $xvy \in L$. Conversely, $xvy \in L$ implies $xuy \in L$. Hence $u \equiv v(P_L)$.

Since $|\{f_u \mid u \in X^*\}| \leq |S|^{|S|}$, the number of congruence classes of P_L is less than or equal to $|S|^{|S|}$.

($\Leftarrow$) Let $u \in X^*$ and let $[u] = \{v \in X^* \mid u \equiv v(P_L)\}$. For any $u \in X^*$ and $a \in X$, we define δ as $\delta([u], a) = [ua]$. Since $[u] = [v]$ implies $[ua] = [va]$ for $u, v \in X^*$ and $a \in X$, δ is well defined as a function of $S \times X$ into S where $S = \{[u] \mid u \in X^*\}$. Notice that S is finite. Thus we can construct the finite acceptor $\boldsymbol{A} = (S, X, \delta, [\epsilon], \{[u] \mid u \in L\})$.

We show that $L = \mathcal{T}(\boldsymbol{A})$. Let $u \in L$. Then $\delta([\epsilon], u) = [u]$ and hence $u \in \mathcal{T}(\boldsymbol{A})$. Therefore, $L \subseteq \mathcal{T}(\boldsymbol{A})$.

Let $u \in \mathcal{T}(\boldsymbol{A})$. Then $\delta([\epsilon], u) = [u] \in \{[u] \mid u \in L\}$ and hence $u \in L$ where we notice that $v \in L$ if $u \in L$ and $[u] = [v]$. Thus $\mathcal{T}(\boldsymbol{A}) \subseteq L$ and $L = \mathcal{T}(\boldsymbol{A}) \subseteq L$. This completes the proof of the proposition.

The following problems are decidable.

Proposition 4.1.4 *(1) For a given finite acceptor (nondeterministic acceptor)* $\boldsymbol{A}$*, it is decidable whether* $\mathcal{T}(\boldsymbol{A}) = \emptyset$*,* $|\mathcal{T}(\boldsymbol{A})| < \infty$ *or* $|\mathcal{T}(\boldsymbol{A})| = \infty$*. (2) For given two finite acceptors (nondeterministic acceptors)* $\boldsymbol{A}, \boldsymbol{B}$*, it is decidable whether* $\mathcal{T}(\boldsymbol{A}) = \mathcal{T}(\boldsymbol{B})$*.*

Proof (1) Let $\boldsymbol{A} = (S, X, \delta, s_0, F)$ be a finite acceptor. Assume $\mathcal{T}(\boldsymbol{A}) \neq \emptyset$. Let $u \in X^*$ be one of the shortest words with $u \in \mathcal{T}(\boldsymbol{A})$. Suppose $|u| \geq |S|$. Then $u = a_0a_1a_2\cdots a_k$ where $a_0 = \epsilon, a_1, a_2, \ldots, a_k \in X$. Consider $\delta(s_0, a_0), \delta(s_0, a_0a_1), \delta(s_0, a_0a_1a_2), \ldots, \delta(s_0, a_0a_1a_2\cdots a_k)$. Since $k \geq |S|$, there exist nonnegative integers $i, j = 0, 1, \ldots, k$ such that $\delta(s_0, a_0a_1\cdots a_i) = \delta(s_0, a_0a_1\cdots a_i\cdots a_j)$. Let $u' = a_0a_1a_2\cdots a_ia_{j+1}\cdots a_k$. Then $\delta(s_0, u) = \delta(s_0, u')$. Hence $u' \in \mathcal{T}(\boldsymbol{A})$ and $|u'| = |u| - (j - i) < |u|$. This contradicts the minimality of $|u|$. Therefore, $|u| < |S|$. This result provides the following algorithm to decide whether $\mathcal{T}(\boldsymbol{A}) = \emptyset$: (1) For any $u \in \bigcup_{i<|S|} X^i$, check whether $u \in \mathcal{T}(\boldsymbol{A})$. (2) If there exists $u \in \bigcup_{i<|S|} X^i$ such that $u \in \mathcal{T}(\boldsymbol{A})$, then $\mathcal{T}(\boldsymbol{A}) \neq \emptyset$. Otherwise, $\mathcal{T}(\boldsymbol{A}) = \emptyset$.

Now assume $|\mathcal{T}(\boldsymbol{A})| = \infty$. Let $u \in X^*$ be one of the shortest words with $|u| \geq 2|S|$. Then $u = a_0a_1a_2\cdots a_k$ where $a_0 = \epsilon, a_1, a_2, \ldots, a_k \in X$. By the preceding method, there exist nonnegative integers $i, j = 0, 1, 2, \ldots, |S|$ such that $\delta(s_0, a_0a_1 \cdots a_i) = \delta(s_0, a_0a_1 \cdots a_i \cdots a_j)$. Let $u' = a_0a_1 \cdots a_ia_{j+1} \cdots a_k$. Then $\delta(s_0, u') = \delta(s_0, u) \in F$ and $u' \in \mathcal{T}(\boldsymbol{A})$. By the minimality of $|u|$ and $0 < j-i \leq |S|$, we have $|S| \leq |u'| < 2|S|$. On the other hand, assume that there exists $v \in \mathcal{T}(\boldsymbol{A})$ such that $|S| \leq |v| < 2|S|$. Then $v = b_0b_1b_2 \cdots b_t$ where $b_0 = \epsilon, b_1, b_2, \ldots, b_t \in X$. Then there exist non-negative integers $p, q, 0 \leq p < q \leq |S|$ such that $\delta(s_0, b_0b_1 \cdots b_p) = \delta(s_0, b_0b_1 \cdots b_p \cdots b_q)$. Let $x = b_0b_1 \cdots b_p, y = b_{p+1}b_{p+2} \cdots b_q, z = b_{q+1} \cdots b_t$. Then $v = xyz$ and $xy^iz \in \mathcal{T}(\boldsymbol{A})$ for any $i, i \geq 1$ because $\delta(s_0, xy^iz) = \delta(s_0, xyz) \in F$. Notice that $|y| \geq 1$. Hence $|\mathcal{T}(\boldsymbol{A})| = \infty$. These results provide the following algorithm: (1) For any $u \in \bigcup_{|S| \leq i < 2|S|} X^i$, check whether $u \in \mathcal{T}(\boldsymbol{A})$. (2) If there exists $u \in \bigcup_{|S| \leq i < 2|S|} X^i$ such that $u \in \mathcal{T}(\boldsymbol{A})$, then $|\mathcal{T}(\boldsymbol{A})| = \infty$. Otherwise, $|\mathcal{T}(\boldsymbol{A})| < \infty$.

(2) Let $\boldsymbol{A} = (S, X, \delta, s_0, F)$ and let $\boldsymbol{B} = (T, X, \gamma, t_0, G)$. First consider the finite acceptor $\boldsymbol{A}' = (S, X, \delta, s_0, S \setminus F)$. It is easy to see that $\mathcal{T}(\boldsymbol{A}') = X^* \setminus \mathcal{T}(\boldsymbol{A})$. Now we construct the acceptor $\boldsymbol{A} \times \boldsymbol{B} = (S \times T, \delta \times \gamma, (s_0, t_0), F \times G)$ where $\delta \times \gamma((s, t), a) = (\delta(s, a), \gamma(t, a))$ for any $(s, t) \in S \times T$ and $a \in X$. Let $u \in X^*$. Then it can be easily seen that $u \in \mathcal{T}(\boldsymbol{A} \times \boldsymbol{B})$ if and only if $u \in \mathcal{T}(\boldsymbol{A}) \cap \mathcal{T}(\boldsymbol{B})$. Notice that $\mathcal{T}(\boldsymbol{A}) \setminus \mathcal{T}(\boldsymbol{B}) = \mathcal{T}(\boldsymbol{A}) \cap (X^* \setminus \mathcal{T}(\boldsymbol{B}))$ and $\mathcal{T}(\boldsymbol{B}) \setminus \mathcal{T}(\boldsymbol{A}) = \mathcal{T}(\boldsymbol{B}) \cap (X^* \setminus \mathcal{T}(\boldsymbol{A}))$. Moreover, $\mathcal{T}(\boldsymbol{A}) = \mathcal{T}(\boldsymbol{B})$ if and only if $\mathcal{T}(\boldsymbol{A}) \setminus \mathcal{T}(\boldsymbol{B}) = \emptyset$ and $\mathcal{T}(\boldsymbol{B}) \setminus \mathcal{T}(\boldsymbol{A}) = \emptyset$. These results provide an algorithm to decide whether $\mathcal{T}(\boldsymbol{A}) = \mathcal{T}(\boldsymbol{B})$, i.e. it is decidable whether $\mathcal{T}(\boldsymbol{A}) = \mathcal{T}(\boldsymbol{B})$ for given finite acceptors $\boldsymbol{A}$ and $\boldsymbol{B}$.

Definition 4.1.4 A *pushdown acceptor* $\boldsymbol{A} = (S, X, \Gamma, \delta, s_0, \gamma_0)$ is defined as follows: (1) S is a state set. (2) X is a finite alphabet. (3) Γ is a finite nonempty set consisting of symbols, called *stack symbols*. (4) $s_0 \in S$ and $\gamma_0 \in \Gamma$. (5) δ is the relation satisfying the following property: $\delta(s, a, \gamma) \subseteq S \times \Gamma^*$ for any $s \in S, \gamma \in \Gamma$ and $a \in X \cup \{\epsilon\}$.

Then the computation by $\boldsymbol{A}$ is performed as follows: $(au, s, \alpha\gamma) \vdash_A (u, t, \alpha\gamma')$ where $a \in X \cup \{\epsilon\}, u \in X^*, s, t \in S, \gamma \in \Gamma, \alpha, \gamma' \in \Gamma^*$ and $(t, \gamma') \in \delta(s, a, \gamma)$.

Moreover, the language $\mathcal{N}(\boldsymbol{A})$ accepted by a pushdown acceptor $\boldsymbol{A}$ is $\{u \in X^* \mid (u, s_0, \gamma_0) \vdash_A^* (\epsilon, t, \epsilon)\}$ where $t \in S$ and $\vdash_A^*$ is the transitive closure of $\vdash_A$.

Example 4.1.1 The language $\{a^nb^n \mid n \geq 1\}$ can be accepted by the following pushdown acceptor $\boldsymbol{A} = (\{s_0, s_1\}, \{a, b\}, \{\gamma_0, +\}, \delta, s_0, \gamma_0)$: (1) $\delta(s_0, a, \gamma_0) = (s_0, +)$. (2) $\delta(s_0, b, \gamma_0) = \emptyset$. (3) $\delta(s_0, a, +) = (s_0, ++)$. (4) $\delta(s_0, b, +) = (s_1, \epsilon)$. (5) $\delta(s_1, b, +) = (s_1, \epsilon)$. (6) $\delta(s_1, a, +) = \emptyset$.

Proposition 4.1.5 *A language accepted by a pushdown acceptor is a context-free language. Conversely, a context-free language over X is accepted by a pushdown acceptor.*

Outline of the proof Let $\boldsymbol{A} = (S, X, \Gamma, \delta, s_0, \gamma_0)$ be a pushdown acceptor and let $L = \mathcal{T}(\boldsymbol{A})$. We construct the context-free grammar $G = (V, X, P, S)$ as follows: (1) $V = \{[s, \gamma, t] \mid s, t \in S, \gamma \in \Gamma\} \cup \{T\}$ where T is a new symbol. (2) (2.1) $\{T \to [s_0, \gamma_0, s] \mid s \in S\} \subseteq P$. (2.2) $\{[s, \gamma, s_1] \to a[s_1, \gamma_k, s_2][s_2, \gamma_{k-1}, s_3] \cdots [s_k, \gamma_1, s_{k+1}] \mid a \in X \cup \{\epsilon\}, k = 0, 1, 2, \ldots, \gamma, \gamma_1, \gamma_2, \ldots, \gamma_k \in \Gamma, s, s_1, s_2, \ldots, s_{k+1} \in S, (s_1, \gamma_1\gamma_2 \cdots \gamma_k) \in \delta(s, a, \gamma)\} \subseteq P$. Notice that $[s, \gamma, s_1] \to a$ if $k = 0$, i.e. $(s_1, \epsilon) \in \delta(s, a, \gamma)$.

Then it can be shown that $(u, s_0, \gamma_0) \vdash_A^* (\epsilon, t, \epsilon), u \in X^*, t \in S$ if and only if $T \Rightarrow [s_0, \gamma_0, s] \Rightarrow^* u, s \in S$. Thus $L = \mathcal{L}(\mathcal{G})$.

Conversely, let $\mathcal{G} = (V, X, P, S)$ be a context-free grammar in the Greibach normal form (see Remark 4.1.2) and let $L = \mathcal{L}(\mathcal{G})$. We construct a pushdown acceptor $\boldsymbol{B} = (\{s\}, X, V, \delta, s, S)$ as follows: (1) $\delta(s, a, A) = \{(s, A_kA_{k-1} \cdots A_1) \mid (A \to aA_1A_2 \cdots A_k) \in P, k = 1, 2, \ldots, A_1, A_2, \ldots, A_k \in V\} \cup \{(s, \epsilon) \mid (A \to a) \in P\}$ for $a \in X, A \in V$. (2) $\delta(s, \epsilon, S) = \{(s, \epsilon)\}$ if $(S \to \epsilon) \in P$ and $\delta(s, \epsilon, S) = \emptyset$, otherwise.

Then it is obvious that $S \Rightarrow^* u, u \in X^*$ if and only if $(u, s, S) \vdash_B^* (\epsilon, s, \epsilon)$. Hence $L = \mathcal{T}(\boldsymbol{B})$.

Regarding regular languages and context-free languages, we have the following iteration properties:

Proposition 4.1.6 *(1) Let $L \subseteq X^*$ be a regular language. Then there exists a positive integer $n(L)$ determined only by L satisfying the following property: If $u \in L$ and $|u| \geq n(L)$, then u can be decomposed into $u = vwx$ and $vw^*x \subseteq L$ where $w \in X^+, |w| \leq n(L)$ and $v, x \in X^*$. (2) Let $L \subseteq X^*$ be a context-free language. Then* *there exists a positive integer* $n(L)$ *determined only by* L *satisfying the following property: If* $u \in L$ *and* $|u| \geq n(L)$*, then* u *can be decomposed into* $u = vwxyz$ *and* $\{vw^ixy^iz \mid i \geq 0\} \subseteq L$ *where* $v, w, x, y, z \in X^*, wy \in X^+$ *and* $|wxy| \leq n(L)$.

Proof (1) Let $L \subseteq X^*$ be a regular language and let $\boldsymbol{A} = (S, X, \delta, s_0, F)$ be a finite acceptor with $L = \mathcal{T}(\boldsymbol{A})$. Let $n(L) = |S|$. By the proof of Proposition 4.1.4 (1), $u = vwx, v, w \in X^*, 0 < |w| \leq n(L)$ and $vw^*x \subseteq L$, if $u \in L$ and $|u| \geq n(L)$.

(2) Let $L \subseteq X^*$ be a context-free language and let $\mathcal{G} = (V, X, P, S)$ be a context-free grammar with $L = \mathcal{L}(\mathcal{G})$. Moreover, let $p = |V|$ and let $q = max\{|\beta| \mid \alpha \to \beta \in P\}$. Put $n(L) = q^p + 1$. If $u \in L$ and $|u| \geq n(L)$, then we have the derivation: $S \Rightarrow^* vTz \Rightarrow^* vwTyz \Rightarrow^* vwxyz = u$ where $v, w, x, y, z \in X^*, |x| \geq 1$ and $T \in V$. In the above, $T \Rightarrow^* wTz$ and $T \Rightarrow^* x$ are used. Notice that we can assume that $0 < |wy| \leq n(L)$. Since $T \Rightarrow^* w^iTy^i$ and $T \Rightarrow^* x$ hold for any $i \geq 0$, we have the derivation $S \Rightarrow^* vTz \Rightarrow^* vw^ixy^iz$ for any $i \geq 1$. Hence $u = vwxyz$ and $vw^ixy^iz \in L$ for any $i \geq 0$.

Remark 4.1.4 For regular languages, we can show the following iteration property: *Let* $L \subseteq X^*$ *be a regular language. Then there exists a positive integer* $n(L)$ *determined only by* L *satisfying the following property: If* $u = v^n \in L$ *holds for some* $v \in X^+$ *and* $n \geq n(L)$*, then there exists a positive integer* $p \leq n(L)$ *such that* $v^{n+ip} \in L$ *holds for any* $i \geq 0$.

It seems that the same kind of iteration property holds true for context-free languages. However, the following example shows that the situation is very different for context-free languages.

Let $a, b \in X, a \neq b$ and let $\mathcal{G} = (X, V, P, S)$ be the following context-free grammar where $V = \{S, T, B\}, P = \{S \to aTB, T \to aTB, T \to b, B \to aB, B \to ab\}$. Let $L = \mathcal{L}(\mathcal{G})$. We will show that $(a^kb)^+ \cap L = (a^kb)^{k+1}$ for any $k \geq 1$.

For any $k \geq 1$, the derivation $S \Rightarrow^* a^kbB^k$ is uniquely determined. After that, for any $(i_1, i_2, \ldots, i_k) \in N^k$, we have $a^kbB^k \Rightarrow^* a^kba^{i_1}ba^{i_2}b \cdots a^{i_k}b$. Thus $(a^kb)^+ \cap L = (a^kb)^{k+1}$. Consequently, for any positive integer $p, (a^pb)^{p+1} \in L$ but $(a^pb)^n \notin L$ for any $n > p + 1$. Therefore, the above kind of iteration property does not hold for context-free languages.

For the details of context-free languages and pushdown acceptors, see [24] and [62].

Now, we introduce a few closure properties:

Proposition 4.1.7 *(1) The union of two regular (context-free) languages is regular (context-free). (2) The intersection of two regular languages is regular. (3) The intersection of two context-free languages is not necessarily context-free. (4) The intersection of a regular language and a context-free language is context-free.*

Proof (1) Let $L_i \subseteq X^*, i = 1,2$ be regular (context-free) languages and let $\mathcal{G}_i = (V_i, X, P_i, S_i), i = 1,2$ be regular (context-free) grammars such that $L_i = \mathcal{L}(\mathcal{G}_i), i = 1,2$. We can assume that $V_1 \cap V_2 = \emptyset$. We construct the following grammars: $\mathcal{G} = (V_1 \cup V_2 \cup \{S\}, X, P_1 \cup P_2 \cup \{S \to S_1, S \to S_2\}, S)$ where S is a new variable with $S \notin V_1 \cup V_2$. Then it is obvious that $\mathcal{G}$ is a regular (context-free) grammar and $\mathcal{L}(\mathcal{G}) = \mathcal{L}(\mathcal{G}_1) \cup \mathcal{L}(\mathcal{G}_2) = L_1 \cup L_2$.

(2) Obvious from the proof of Proposition 4.1.4 (2).

(3) Let $X = \{a, b, \ldots\}$. Then $L_1 = \{a^n b^n a^m \mid n, m \geq 1\}$ and $L_2 = \{a^n b^m a^m \mid n, m \geq 1\}$ are context-free. However, $L_1 \cap L_2 = \{a^n b^n a^n \mid n \geq 1\}$ is not context-free (use the iteration property for context-free languages in Proposition 4.1.6).

(4) Let $L \subseteq X^*$ be a regular language and let $M \subseteq X^*$ be a context-free language. Then there exist a finite acceptor $\boldsymbol{A} = (S, X, \delta, s_0, F)$ and a pushdown acceptor $\boldsymbol{B} = (T, X, \Gamma, \theta, t_0, \gamma_0)$ such that $L = \mathcal{T}(\boldsymbol{A})$ and $M = \mathcal{T}(\boldsymbol{B})$. Consider the following pushdown acceptor $\boldsymbol{A} \times \boldsymbol{B} = ((S \times T) \cup \{u_0\}, X, \delta \times \theta, \Gamma \cup \{\#\}, u_0, \gamma_0)$: (1) $\# \notin \Gamma$, (2) $u_0 \notin S \times T$, (3) $\delta \times \theta(u_0, \epsilon, \gamma_0) = \{((s_0, t_0), \#\gamma_0)\}$, (4) $\delta \times \theta((s,t), a, \gamma) = \{((\delta(s,a), t'), \gamma') \mid (t', \gamma') \in \theta(t, a, \gamma)\}$ for any $(s,t) \in S \times T, \gamma \in \Gamma$ and $a \in X \cup \{\epsilon\}$, (4) $\delta \times \theta((s,t), \epsilon, \#) = \{((s,t), \epsilon)\}$ if $s \in F$. Then it is obvious that $u \in \mathcal{T}(\boldsymbol{A} \times \boldsymbol{B})$ if and only if $u \in \mathcal{T}(\boldsymbol{A}) \cap \mathcal{T}(\boldsymbol{B})$. Hence $L \cap M$ is context-free.

Remark 4.1.5 The following two results can be easily verified: (1) Let $A, B \subseteq X^*$. Then the language $AB = \{uv \mid u \in A, v \in B\}$ is called the *concatenation* of A and B. Then the concatenation of two regular (context-free) languages is regular (context-free). (2)

Let X, Y be finite alphabets and let ρ be a mapping of X^* into Y^*. Then ρ is called a *homomorphism* of X^* into Y^* if $\rho(uv) = \rho(u)\rho(v)$ holds for any $u, v \in X^*$. Then the image of a regular (context-free) language by a homomorphism is regular (context-free).

4.2 Operations on languages

There are some basic operations on languages related to computation. Especially, the shuffle operation can be regarded as a mathematical model of parallel computation and it has been intensively studied in formal language theory. For instance, some types of regular expressions of shuffle operators were dealt with in [1], [2], [49] and [50] - [53]. A constrained form of shuffle product, i.e. the literal shuffle is defined in [7], while a special kind of literal shuffle product of a language is studied in [46]. In this book, many problems related to shuffle operations will be discussed.

First we define the n-insertion and shuffle product of words and languages.

Definition 4.2.1 Let $u, v \in X^*$ and let n be a positive integer. Then the *n-insertion* of u into v, denoted by $u \rhd^{[n]} v$, is $\{v_1u_1v_2u_2 \cdots v_nu_nv_{n+1} \mid u = u_1u_2\cdots u_n,\ u_1, u_2, \ldots, u_n \in X^*,\ v = v_1v_2\cdots v_nv_{n+1},\ v_1, v_2, \ldots, v_n, v_{n+1} \in X^*\}$. For languages $A, B \subseteq X^*$, the *n-insertion* $A \rhd^{[n]} B$ of A into B is $\bigcup_{u\in A, v\in B} u \rhd^{[n]} v$. For $u, v \in X^*$, the *shuffle product* $u \diamond v$ is defined as $\bigcup_{n\geq 1} u \rhd^{[n]} v$. Moreover, the *shuffle product* $A \diamond B$ of languages $A, B \subseteq X^*$ is defined as $\bigcup_{n\geq 1} A \rhd^{[n]} B$.

In Section 4.3, we will provide some properties of n-insertions. For instance, the n-insertion of a regular language into a regular language is regular but the n-insertion of a context-free language into a context-free language is not always context-free. However, it can be shown that the n-insertion of a regular (context-free) language into a context-free (regular) language is context-free. In Section 4.4, we prove that, for a given regular language $L \subseteq X^*$ and a positive integr n, it is decidable whether $L = A \rhd^{[n]} B$ for some nontrivial regular languages $A, B \subseteq X^*$. Here a language $C \subseteq X^*$ is said to be *nontrivial* if $C \neq \{\epsilon\}$.

4.3 Shuffle products and n-insertions

First, we consider the shuffle product of languages.

Lemma 4.3.1 *Let $A, B \subseteq X^*$ be regular languages. Then $A \diamond B$ is a regular language.*

Proof By $\overline{X}$ we denote the new alphabet $\{\overline{a} \mid a \in X\}$. Let $\boldsymbol{A} = (S, X, \delta, s_0, F)$ be a finite deterministic acceptor with $\mathcal{T}(\boldsymbol{A}) = A$ and let $\boldsymbol{B} = (T, X, \theta, t_0, G)$ be a finite acceptor with $\mathcal{T}(\boldsymbol{B}) = B$. Define the acceptor $\overline{\boldsymbol{B}} = (T, \overline{X}, \overline{\theta}, t_0, G)$ as $\overline{\theta}(t, \overline{a}) = \theta(t, a)$ for any $t \in T$ and $a \in X$. Let ρ be the homomorphism of $(X \cup \overline{X})^*$ onto X^* defined as $\rho(a) = \rho(\overline{a}) = a$ for any $a \in X$. Moreover, let $\mathcal{T}(\overline{\boldsymbol{B}}) = \overline{B}$. Then $\rho(\overline{B}) = \{\rho(\overline{u}) \mid \overline{u} \in \overline{B}\} = B$ and $\rho(A \diamond \overline{B}) = A \diamond B$. Hence, to prove the lemma, it is enough to show that $A \diamond \overline{B}$ is a regular language over $X \cup \overline{X}$. Consider the acceptor $\boldsymbol{A} \diamond \overline{\boldsymbol{B}} = (S \times T, X \cup \overline{X}, \delta \diamond \overline{\theta}, (s_0, t_0), F \times G)$ where $\delta \diamond \overline{\theta}((s,t), a) = (\delta(s,a), t)$ and $\delta \diamond \overline{\theta}((s,t), \overline{a}) = (s, \theta(t,a))$ for any $(s,t) \in S \times T$ and $a \in X$. Then it is easy to see that $w \in \mathcal{T}(\boldsymbol{A} \diamond \overline{\boldsymbol{B}})$ if and only if $w \in A \diamond \overline{B}$, i.e. $A \diamond \overline{B}$ is regular. This completes the proof of the lemma.

Proposition 4.3.1 *Let $A, B \subseteq X^*$ be regular languages and let n be a positive integer. Then $A \rhd^{[n]} B$ is a regular language.*

Proof Let the notations of $\overline{X}$, $\overline{B}$ and ρ be the same as in Lemma 4.3.1. Notice that $A \rhd^{[n]} \overline{B} = (A \diamond \overline{B}) \cap (\overline{X}^* X^*)^n \overline{X}^*$. Since $(\overline{X}^* X^*)^n \overline{X}^*$ is regular, $A \rhd^{[n]} \overline{B}$ is regular. Consequently, $A \rhd^{[n]} B = \rho(A \rhd^{[n]} \overline{B})$ is regular.

Remark 4.3.1 We show that the n-insertion of a context-free language into a context-free language is not always context-free. For instance, $A = \{a^n b^n \mid n \geq 1\}$ and $B = \{c^n d^n \mid n \geq 1\}$ are context-free languages over $\{a, b\}$ and $\{c, d\}$, respectively. However, since $(A \rhd^{[2]} B) \cap a^+ c^+ b^+ d^+ = \{a^n c^m b^n d^m \mid n, m \geq 1\}$ is not context-free (use Proposition 4.1.6), $A \rhd^{[2]} B$ is not context-free. However, as will be shown in Corollary 4.3.1, $A \rhd^{[1]} B$ is a context-free language for any context-free languages A and B (see also [54]). Usually, a 1-insertion is called an *insertion*.

Now consider the shuffle product of a regular language and a context-free language.

Proposition 4.3.2 *Let $A \subseteq X^*$ be a regular language and let $B \subseteq X^*$ be a context-free language. Then $A \diamond B$ is a context-free language.*

Proof The notations which we will use for the proof are assumed to be the same as the previous cases. Let $\boldsymbol{A} = (S, X, \delta, s_0, F)$ be a finite acceptor with $\mathcal{T}(\boldsymbol{A}) = A$ and let $\boldsymbol{B} = (T, X, \Gamma, \theta, t_0, \gamma_0)$ be a pushdown acceptor with $\mathcal{N}(\boldsymbol{B}) = B$. Let $\overline{\boldsymbol{B}} = (T, \overline{X}, \Gamma, \overline{\theta}, t_0, \gamma_0)$ be a pushdown acceptor such that $\overline{\theta}(t, \overline{a}, \gamma) = \theta(t, a, \gamma)$ for any $t \in T, a \in X \cup \{\epsilon\}$ and $\gamma \in \Gamma$. Then $\rho(\mathcal{N}(\overline{\boldsymbol{B}})) = B$. Now define the pushdown acceptor $\boldsymbol{A} \diamond \overline{\boldsymbol{B}} = (S \times T, X \cup \overline{X}, \Gamma \cup \{\#\}, \delta \diamond \overline{\theta}, (s_0, t_0), \gamma_0)$ as follows: (1) $\forall a \in X \cup \{\epsilon\}, \delta \diamond \overline{\theta}((s_0, t_0), a, \gamma_0) = \{((\delta(s_0, a), t_0), \#\gamma_0)\}$, $\delta \diamond \overline{\theta}((s_0, t_0), \overline{a}, \gamma_0) = \{((s_0, t'), \#\gamma') \mid (t', \gamma') \in \overline{\theta}(t_0, \overline{a}, \gamma_0)\}$. (2) $\forall a \in X, \forall (s, t) \in S \times T, \forall \gamma \in \Gamma \cup \{\#\}, \delta \diamond \overline{\theta}((s, t), a, \gamma) = \{((\delta(s, a), t), \gamma)\}$. (3) $\forall a \in X \cup \{\epsilon\}, \forall (s, t) \in S \times T, \forall \gamma \in \Gamma, \delta \diamond \overline{\theta}((s, t), \overline{a}, \gamma) = \{((s, t'), \gamma') \mid (t', \gamma') \in \overline{\theta}(t, \overline{a}, \gamma)\}$. (4) $\forall (s, t) \in F \times T, \delta \diamond \overline{\theta}((s, t), \epsilon, \#) = \{((s, t), \epsilon)\}$.

Let $w = \overline{v}_1 u_1 \overline{v}_2 u_2 \cdots \overline{v}_n u_n \overline{v}_{n+1}$ where $u_1, u_2, \ldots, u_n \in X^*$ and $\overline{v}_1, \overline{v}_2, \ldots, \overline{v}_{n+1} \in \overline{X}^*$. Assume $\delta \diamond \overline{\theta}((s_0, t_0), w, \gamma_0) \neq \emptyset$. Then we have the following computation: $((s_0, t_0), w, \gamma_0) \vdash^*_{A \diamond \overline{B}} ((\delta(s_0, u_1 u_2 \cdots u_n), t'), \epsilon, \# \cdots \#\gamma')$ where $(t', \gamma') \in \overline{\theta}(t_0, \overline{v}_1 \overline{v}_2 \cdots \overline{v}_{n+1}, \gamma_0)$. If $w \in \mathcal{N}(\boldsymbol{A} \diamond \overline{\boldsymbol{B}})$, then $(\delta(s_0, u_1 u_2 \cdots u_n), t'), \epsilon, \# \cdots \#\gamma') \vdash^*_{A \diamond \overline{B}} (\delta(s_0, u_1 u_2 \cdots u_n), t'), \epsilon, \epsilon)$. Hence $(\delta(s_0, u_1 u_2 \cdots u_n), t') \in F \times T$ and $\gamma' = \epsilon$. This means that $u_1 u_2 \cdots u_n \in A$ and $\overline{v}_1, \overline{v}_2, \ldots, \overline{v}_{n+1} \in \overline{B}$. Therefore, $w \in A \times \overline{B}$. Now let $w \in A \times \overline{B}$. Then by the above configuration, we have $((s_0, t_0), w, \gamma_0) \vdash^*_{A \diamond \overline{B}} ((\delta(s_0, u_1 u_2 \cdots u_n), t'), \epsilon, \# \cdots \#) \vdash^*_{A \diamond \overline{B}} ((\delta(s_0, u_1 u_2 \cdots u_n), t'), \epsilon, \epsilon)$ and $w \in \mathcal{N}(\boldsymbol{A} \diamond \overline{\boldsymbol{B}})$. Thus $A \diamond \overline{B} = \mathcal{N}(\boldsymbol{A} \diamond \overline{\boldsymbol{B}})$ and $A \diamond \overline{B}$ is context-free. Since $\rho(A \diamond \overline{B}) = A \diamond B$, $A \diamond B$ is context-free.

Regarding the n-insertion of a regular (context-free) language into a context-free (regular) language, we have:

Proposition 4.3.3 *Let $A \subseteq X^*$ be a regular (context-free) language and let $B \subseteq X^*$ be a context-free (regular) language. Then $A \triangleright^{[n]} B$ is a context-free language.*

Proof We consider the case that $A \subseteq X^*$ is regular and $B \subseteq X^*$ is context-free. Since $A \rhd^{[n]} \overline{B} = (A \diamond \overline{B}) \cap (\overline{X}^* X^*)^n \overline{X}^*$ and $(\overline{X}^* X^*)^n \overline{X}^*$ is regular, $A \rhd^{[n]} \overline{B}$ is context-free. Consequently, $A \rhd^{[n]} B = \rho(A \rhd^{[n]} \overline{B})$ is context-free.

Corollary 4.3.1 *Let $A, B \subseteq X^*$ be context-free language. Then $A \rhd^{[1]} B$ is context-free.*

Proof Let $\mathcal{G}_A = (V_A, X, P_A, S_A)$ be a context-free grammar with $\mathcal{L}(\mathcal{G}_A) = A$. Consider the language $S_A \rhd^{[1]} B$ over $X \cup \{S_A\}$. By the above proposition, $S_A \rhd^{[1]} B$ is a context-free language over $X \cup \{S_A\}$. Let $\mathcal{G}_C = (V_C, X, P_C, S_C)$ be a context-free grammar such that $V_A \cap V_C = \emptyset$ and $\mathcal{L}(\mathcal{G}_C) = S_A \rhd^{[1]} B$. Let $\mathcal{G}_D = (V_C \cup \{S_A\}, P_C \cup P_A, S_C)$. Then $\mathcal{G}_D$ is a context-free grammar and $\mathcal{L}(\mathcal{G}_A) = A \rhd^{[1]} B$. This completes the proof of the corollary.

4.4 Decompositions

Let $L \subseteq X^*$ be a regular language and let $\mathbf{A} = (S, X, \delta, s_0, F)$ be a finite acceptor accepting the language L, i.e. $\mathcal{T}(\mathbf{A}) = L$. For $u, v \in X^*$, by $u \sim v$ we denote the equivalence relation of finite index on X^* such that $\delta(s, u) = \delta(s, v)$ for any $s \in S$. Then it is obvious that $u \equiv v(P_L)$ if $u \sim v$. Let $[u] = \{v \in X^* \mid u \sim v\}$ for $u \in X^*$. It is easy to see that $[u]$ can be effectively constructed using $\mathbf{A}$ for any $u \in X^*$. Now let n be a positive integer. We consider the decomposition $L = A \rhd^{[n]} B$. Let $K_n = \{([u_1], [u_2], \ldots, [u_n]) \mid u_1, u_2, \ldots, u_n \in X^*\}$. Notice that K_n is a finite set.

Lemma 4.4.1 *There is an algorithm to construct K_n.*

Proof Obvious from the fact that $[u]$ can be effectively constructed for any $u \in X^*$ and $\{[u] \mid u \in X^*\} = \{[u] \mid u \in X^*, |u| \leq |S|^{|S|}\}$. Here $|u|$ and $|S|$ denote the length of u and the cardinality of S, respectively.

For $u \in X^*$, we define $\rho_n(u)$ by $\{([u_1], [u_2], \ldots, [u_n]) \mid u = u_1 u_2 \cdots u_n, u_1, u_2, \ldots, u_n \in X^*\}$. Moreover, let $\mu = ([u_1], [u_2], \ldots, [u_n]) \in K_n$ and let $B_\mu = \{v \in X^* \mid \forall v = v_1 v_2 \cdots v_n v_{n+1}, v_1, v_2, \cdots .v_n, v_{n+1} \in X^*, \{v_1\}[u_1]\{v_2\}[u_2] \cdots \{v_n\}[u_n]\{v_{n+1}\} \subseteq L\}$.

Lemma 4.4.2 *$B_\mu \subseteq X^*$ is a regular language and it can be effectively constructed.*

Proof Let $i = 0, 1, \ldots, n$, let $S^{(i)} = \{s^{(i)} \mid s \in S\}$ and let $\tilde{S} = \bigcup_{0 \le i \le n} S^{(i)}$. We define the following nondeterministic acceptor $\tilde{\boldsymbol{A}}' = (\tilde{S}, X, \tilde{\delta}, \{s_0^{(0)}\}, S^{(n)} \setminus F^{(n)})$ with ϵ-move where $F^{(n)} = \{s^{(n)} \mid s \in F\}$. Here $\tilde{\delta}$ is defined as follows: $\tilde{\delta}(s^{(i)}, a) = \{\delta(s, a)^{(i)}, \delta(s, au_{i+1})^{(i+1)}\}$ for any $a \in X \cup \{\epsilon\}$ and any $i = 0, 1, \ldots, n-1$ and $\tilde{\delta}(s^{(n)}, a) = \{\delta(s, a)^{(n)}\}$ for any $a \in X$.

Let $v \in \mathcal{T}(\tilde{\boldsymbol{A}}')$. Then $\delta(s_0, v_1u_1v_2u_2 \cdots v_nu_nv_{n+1})^{(n)} \in \tilde{\delta}(s_0^{(0)}, v_1 v_2 \cdots v_nv_{n+1}) \cap (S^{(n)} \setminus F^{(n)})$ for some $v = v_1v_2 \cdots v_nv_{n+1}, v_1, v_2, \ldots, v_n, v_{n+1} \in X^*$. Hence $v_1u_1v_2u_2 \cdots v_nu_nv_{n+1} \notin L$, i.e. $v \in X^* \setminus B_\mu$. Now let $v \in X^* \setminus B_\mu$. Then there exists $v = v_1v_2 \cdots v_nv_{n+1}, v_1, v_2, \ldots, v_n, v_{n+1} \in X^*$ such that $v_1u_1v_2u_1 \cdots v_nu_nv_{n+1} \notin L$. Therefore, $\tilde{\delta}(s_0^{(0)}, v_1v_2 \cdots v_nv_{n+1}) \in S^{(n)} \setminus F^{(n)}$, i.e. $v = v_1v_2 \cdots v_nv_{n+1} \in \mathcal{T}(\tilde{\boldsymbol{A}}')$. Consequently, $B_\mu = X^* \setminus \mathcal{T}(\tilde{\boldsymbol{A}}')$ and B_μ is regular. Notice that $X^* \setminus \mathcal{T}(\tilde{\boldsymbol{A}}')$ can be effectively constructed.

Symmetrically, we will consider $\nu = ([v_1], [v_2], \ldots, [v_n], [v_{n+1}]) \in K_{n+1}$ and $A_\nu = \{u \in X^* \mid \forall u = u_1u_2 \cdots u_n, u_1, u_2, \ldots, u_n \in X^*, [v_1]\{u_1\}[v_2]\{u_2\} \cdots [v_n]\{u_n\}[v_{n+1}] \subseteq L\}$.

Lemma 4.4.3 *$A_\nu \subseteq X^*$ is a regular language and it can be effectively constructed.*

Proof Let $S^{(i)} = \{s^{(i)} \mid s \in S\}, 1 \le i \le n+1$, and let $\overline{S} = \bigcup_{1 \le i \le n+1} S^{(i)}$. Then we define the following nondeterministic acceptor $\overline{\boldsymbol{B}}' = (\overline{S}, X, \overline{\delta}, \{\delta(s_0, v_1)^{(1)}\}, S^{(n+1)} \setminus F^{(n+1)})$ with ϵ-move where $F^{(n+1)} = \{s^{(n+1)} \mid s \in F\}$. The state transition relation $\overline{\delta}$ is defined as follows: $\overline{\delta}(s^{(i)}, a) = \{\delta(s, a)^{(i)}, \delta(s, au_{i+1})^{(i+1)}\}$ for any $a \in X \cup \{\epsilon\}$ and any $i = 1, 2, \ldots, n$.

In the same way as in the proof of Lemma 4.4.2, we can prove that $A_\nu = X^* \setminus \mathcal{T}(\overline{\boldsymbol{B}'})$. Therefore, A_ν is regular. Notice that $X^* \setminus \mathcal{T}(\overline{\boldsymbol{B}'})$ can be effectively constructed.

Proposition 4.4.1 *Let $A, B \subseteq X^*$ and let $L \subseteq X^*$ be a regular language. If $L = A \rhd^{[n]} B$, then there exist regular languages $A', B' \subseteq X^*$ such that $A \subseteq A', B \subseteq B'$ and $L = A' \rhd^{[n]} B'$.*

Proof Put $B' = \bigcap_{\mu \in \rho_n(A)} B_\mu$. Let $v \in B$ and let $\mu \in \rho_n(A)$. Since $\mu \in \rho_n(A)$, there exists $u \in A$ such that $\mu = ([u_1], [u_2], \ldots, [u_n])$ and $u = u_1 u_2 \cdots u_n, u_1, u_2, \ldots, u_n \in X^*$. By $u \rhd^{[n]} v \subseteq L$, we have $\{v_1\}[u_1]\{v_2\}[u_2] \cdots \{v_n\}[u_n]\{v_{n+1}\} \subseteq L$ for any $v = v_1 v_2 \cdots v_n v_{n+1}$, $v_1, v_2, \ldots, v_n, v_{n+1} \in X^*$. This means that $v \in B_\mu$. Thus $B \subseteq \bigcap_{\mu \in \rho_n(A)} B_\mu = B'$. Now assume $u \in A$ and $v \in B'$. Let $u = u_1 u_2 \cdots u_n, u_1, u_2, \ldots, u_n \in X^*$ and let $\mu = ([u_1], [u_2], \ldots, [u_n]) \in \rho_n(u) \subseteq \rho_n(A)$. Since $v \in B' \subseteq B_\mu$, we have $v_1 u_1 v_2 u_2 \cdots v_n u_n v_{n+1} \in \{v_1\}[u_1]\{v_2\}[u_2] \cdots \{v_n\}[u_n]\{v_{n+1}\} \subseteq L$ for any $v = v_1 v_2 \cdots v_n v_{n+1}$, $v_1, v_2, \ldots, v_n, v_{n+1} \in X^*$. Hence $u \rhd^{[n]} v \subseteq L$ and $A \rhd^{[n]} B' \subseteq L$. On the other hand, since $B \subseteq B'$ and $A \rhd^{[n]} B = L$, we have $A \rhd^{[n]} B' = L$. Symmetrically, put $A' = \bigcap_{\nu \in \rho_{n+1}(B')} A_\nu$. By the same way as the above, we can prove that $A \subseteq A'$ and $L = A' \rhd^{[n]} B'$.

From Lemma 4.4.2 and Lemma 4.4.3, it follows that A' and B' are regular.

Theorem 4.4.1 *For any regular language $L \subseteq X^*$ and a positive integer n, it is decidable whether $L = A \rhd^{[n]} B$ for some nontrivial regular languages $A, B \subseteq X^*$.*

Proof Let $\mathbf{A} = \{A_\nu \mid \nu \in K_{n+1}\}$ and let $\mathbf{B} = \{B_\mu \mid \mu \in K_n\}$. By the preceding lemmata, $\mathbf{A}$, $\mathbf{B}$ are finite sets of regular languages which can be effectively constructed. Assume that $L = A \rhd^{[n]} B$ for some nontrivial regular languages $A, B \subseteq X^*$. In this case, by Proposition 4.4.1, there exist regular languages $A \subseteq A'$ and $B \subseteq B'$ which are an intersection of languages in $\mathbf{A}$ and an intersection of languages in $\mathbf{B}$, respectively. It is obvious that A', B' are nontrivial languages. Thus we have the following algorithm: (1) Take any languages from $\mathbf{A}$ and let A' be their intersection. (2) Take any languages from $\mathbf{B}$ and let B' be their intersection. (3) Calculate $A' \rhd^{[n]} B'$. (4) If $A' \rhd^{[n]} B' = L$, then the output is "YES". (5) If the output is "NO", search another pair of $\{A', B'\}$ until obtaining the output "YES". (6) This procedure terminates after a finite-step trial. (7) Once we get the output "YES", then $L = A \rhd^{[n]} B$ for some nontrivial regular languages $A, B \subseteq X^*$. (8) Otherwise, there are no such decompositions.

Let n be a positive integer. By $\mathcal{F}(n,X)$, we denote the class of finite languages $\{L \subseteq X^* \mid max\{|u| \mid u \in L\} \leq n\}$. Then the following result by C. Câmpeanu et al. [14] can be obtained as a corollary of Theorem 4.4.1.

Corollary 4.4.1 *For a given positive integer n and a regular language $A \subseteq X^*$, the problem whether $A = B \diamond C$ for a nontrivial language $B \in \mathcal{F}(n,X)$ and a nontrivial regular language $C \subseteq X^*$ is decidable.*

Proof Obvious from the following fact: If $u, v \in X^*$ and $|u| \leq n$, then $u \diamond v = u \rhd^{[n]} v$.

The proof of the above corollary will be given in a different way in Section 4.5.

Remark 4.4.1 A decompositon $L = A \rhd^{[n]} B$ of a regular language $L \subseteq X^*$ is said to be *maximal* if $L = A \rhd^{[n]} B = A' \rhd^{[n]} B', A \subseteq A', B \subseteq B'$ imply that $A = A'$ and $B = B'$. Following the above method, it can be shown that $A \subseteq X^*, B \subseteq X^*$ are regular for any maximal decomposition $L = A \rhd^{[n]} B$. Moreover, we can obtain all maximal decompositions of a regular language $L \subseteq X^*$.

Remark 4.4.2 Let $u, v \in X^*$ and let n be a positive integer. Then $u \diamond^{[n]} v = \{u_1 v_1 u_2 v_2 \cdots u_n v_n \mid u = u_1 u_2 \cdots u_n, u_1, u_2, \ldots, u_n \in X^*, v = v_1 v_2 \cdots v_n, v_1, v_2, \ldots, v_n \in X^*\}$ is called the *n-shuffle* of u and v and for languages $A, B \subseteq X^*, \bigcup_{u \in A, v \in B} u \diamond^{[n]} v$ is called the *n-shuffle* of A and B. Then the analogous results as in the case of n-insertions (including the decomposition results) can be obtained.

4.5 Equations on languages

In this section, we consider the following problem: Let $A, B \subseteq X^*$ be regular languages. Then can we obtain a solution $C \subseteq X^*$ of the language equation $A = B \diamond C$? Obviously, this problem is equivalent to a partial shuffle decomposition problem for regular languages.

Now let $\boldsymbol{A} = (S, X, \delta, s_0, F)$ be a finite acceptor with $\mathcal{T}(\boldsymbol{A}) = A$ and let $\boldsymbol{B} = (T, X, \gamma, t_0, G)$ be a finite acceptor with $\mathcal{T}(\boldsymbol{B}) = B$.

We will look for a regular language C over X such that $A = B \diamond C$. By $\overline{X}$, we denote the alphabet $\{\overline{a} \mid a \in X\}$ with $X \cap \overline{X} = \emptyset$. Let $\overline{\boldsymbol{B}} = (T, X \cup \overline{X} \cup \{\#\}, \overline{\gamma}, t_0, G)$ where $\overline{\gamma}$ is defined as follows:

For $t \in T$ and $a \in X$, $\overline{\gamma}(t, a) = t$, $\overline{\gamma}(t, \overline{a}) = \gamma(t, a)$. Moreover, $\overline{\gamma}(t, \#) = t$ if $t \in G$.

Then the following can be easily shown.

Fact *Let $a_1a_2 \cdots a_n \in X^*$ where $a_i \in X, i = 1, 2, \ldots, n$. Then $a_1a_2 \cdots a_n \in \mathcal{T}(\boldsymbol{B})$ if and only if $u_1\overline{a}_1u_2\overline{a}_2 \cdots u_n\overline{a}_nu_{n+1}\# \in \mathcal{T}(\overline{\boldsymbol{B}})$ where $u_1, u_2, \ldots, u_n \in X^*$.*

Let $\boldsymbol{A}_1 = (\overline{S}, X \cup \overline{X} \cup \{\#\}, \overline{\delta}, s_0, \{\alpha, \omega\})$ and let $\boldsymbol{A}_2 = (\overline{S}, X \cup \overline{X} \cup \{\#\}, \overline{\delta}, s_0, \{\alpha\})$ where $\overline{S} = (\cup_{a \in X \cup \{\epsilon\}} S^{(a)}) \cup \{\alpha, \omega\}$. Here $S^{(\epsilon)}$ is regarded as S where ϵ is the empty word. For $s \in S, t \in S \setminus F, t' \in F, a \in X \cup \{\epsilon\}, b \in X$ and $\{\#\}$, $\overline{\delta}$ is defined as follows:

$\overline{\delta}(s^{(a)}, b) = \delta(s, b)^{(a)}, \overline{\delta}(s^{(a)}, \overline{b}) = \delta(s, b)^{(b)}, \overline{\delta}(t^{(a)}, \#) = \{\alpha\}$ and $\overline{\delta}(t'^{(a)}, \#) = \{\omega\}$.

We consider the following two acceptors:

$\boldsymbol{C}_1 = (\overline{S} \times T, X \cup \overline{X} \cup \{\#\}, \overline{\delta} \times \overline{\gamma}, (s_0, t_0), \{\alpha, \omega\} \times G)$, $\boldsymbol{C}_2 = (\overline{S} \times T, X \cup \overline{X} \cup \{\#\}, \overline{\delta} \times \overline{\gamma}, (s_0, t_0), \{\alpha\} \times G)$ where $\overline{\delta} \times \overline{\gamma}((\overline{s}, t), a) = (\overline{\delta}(\overline{s}, a), \overline{\gamma}(t, a))$ for $(\overline{s}, t) \in \overline{S} \times T$ and $a \in X$.

Now consider the following homomorphism ρ of $(X \cup \overline{X} \cup \{\#\})^*$ into X^*: $\rho(a) = a, \rho(\overline{a}) = \epsilon$ for $a \in X$ and $\rho(\#) = \epsilon$.

Lemma 4.5.1 *The acceptors accepting the languages $\rho(\mathcal{T}(\boldsymbol{C}_1))$ and $\rho(\mathcal{T}(\boldsymbol{C}_2))$ can be effectively constructed.*

Proof Let $i = 1, 2$. From $\boldsymbol{C}_i$, we can construct a regular grammar $\mathcal{G}_i$ such that $\mathcal{L}(\mathcal{G}_i) = \mathcal{T}(\boldsymbol{C}_i)$ with the production rules of the form $A \to aB$ (A, B are variables and $a \in X \cup \overline{X} \cup \{\#\}$). Replacing every rule of the form $A \to aB$ in $\mathcal{G}_i$ by $A \to \rho(a)B$, we can obtain a new grammar $\mathcal{G}'_i$. Then it is clear that $\rho(\mathcal{T}(\boldsymbol{C}_i)) = \mathcal{L}(\mathcal{G}'_i)$. Using this grammar $\mathcal{G}'_i$, we can construct an acceptor $\boldsymbol{D}_i$ such that $\mathcal{T}(\boldsymbol{D}_i) = \mathcal{L}(\mathcal{G}'_i)$ i.e. $\rho(\mathcal{T}(\boldsymbol{C}_i)) = \mathcal{T}(\boldsymbol{D}_i)$. Notice that all the above procedures are effectively performed. This completes the proof of the lemma.

Proposition 4.5.1 *Let $u \in X^*$. Then $\{u\} \diamond B \subseteq A$ if and only if $u \in \rho(\mathcal{T}(\boldsymbol{C}_1)) \setminus \rho(\mathcal{T}(\boldsymbol{C}_2))$.*

Proof (⇒) Let $u = u_1u_2 \cdots u_nu_{n+1} \in X^*$ and let $a_1a_2 \cdots a_n \in B$ where $u_1, u_2, \ldots, u_n, u_{n+1} \in X^*$ and $a_1, a_2, \ldots, a_n \in X$. Then we have $\overline{\delta} \times \overline{\gamma}((s_0, t_0), u_1\overline{a}_1u_2\overline{a}_2 \cdots u_n\overline{a}_nu_{n+1}\#) = (\overline{\delta}(s_0, u_1\overline{a}_1u_2\overline{a}_2 \cdots u_n \overline{a}_nu_{n+1}\#), \overline{\gamma}(t_0, u_1\overline{a}_1u_2\overline{a}_2 \cdots u_n\overline{a}_nu_{n+1}\#)) = (\overline{\delta}(\delta(s_0, u_1a_1u_2a_2 \cdots u_n a_nu_{n+1})^{(a_n)}, \#), \overline{\gamma}(\gamma(s_0, a_1a_2 \cdots a_n)^{(a_n)}, \#)) = (\omega, \gamma(t_0, a_1a_2 \cdots a_n)) \in \{\omega\} \times G$. Therefore, $u_1\overline{a}_1u_2\overline{a}_2 \cdots u_n\overline{a}_nu_{n+1}\# \in \mathcal{T}(\boldsymbol{C}_1) \setminus \mathcal{T}(\boldsymbol{C}_2)$. Thus $u = u_1u_2 \cdots u_nu_{n+1} = \rho(u_1\overline{a}_1u_2\overline{a}_2 \cdots u_n\overline{a}_nu_{n+1}\#) \in \rho(\mathcal{T}(\boldsymbol{C}_1)) \setminus\rho(\mathcal{T}(\boldsymbol{C}_2))$.

(⇐) Suppose that $\{u\}\diamond B \subseteq A$ does not hold though $u \in \rho(\mathcal{T}(\boldsymbol{C}_1)) \setminus\rho(\mathcal{T}(\boldsymbol{C}_2))$. Then there exist $u = u_1u_2 \cdots u_nu_{n+1} \in X^*$ and $a_1a_2 \cdots a_n \in B$ such that $u_1a_1u_2a_2 \cdots u_na_nu_{n+1} \notin A$. Thus $\overline{\gamma}(t_0, u_1\overline{a}_1u_2\overline{a}_2 \cdots u_n\overline{a}_nu_{n+1}\#) = \overline{\gamma}(\gamma(t_0, a_1a_2 \cdots a_n), \#) = \gamma(t_0, a_1a_2 \cdots a_n) \in G$. On the other hand, since $u_1a_1u_2a_2 \cdots u_na_nu_{n+1} \notin A$, we have $\overline{\delta}(s_0, u_1\overline{a}_1 u_2\overline{a}_2 \cdots u_n\overline{a}_nu_{n+1}\#) = \overline{\delta}(\delta(s_0, u_1a_1u_2a_2 \cdots u_na_nu_{n+1})^{(a_n)}, \#) = \{\alpha\}$. Hence $\overline{\delta} \times \overline{\gamma}((s_0, t_0), u_1\overline{a}_1u_2\overline{a}_2 \cdots u_n\overline{a}_nu_{n+1}\#) \in \{\alpha\} \times G$, i.e. $u_1\overline{a}_1u_2 \overline{a}_2 \cdots u_n\overline{a}_nu_{n+1}\# \in \mathcal{T}(\boldsymbol{C}_2)$. Therefore, $u = \rho(u_1\overline{a}_1u_2\overline{a}_2 \cdots u_n\overline{a}_nu_{n+1} \#) \in \rho(\mathcal{T}(\boldsymbol{C}_2))$. On the other hand, it is obvious that $u_1\overline{a}_1u_2\overline{a}_2 \cdots u_n \overline{a}_nu_{n+1}\# \in \mathcal{T}(\boldsymbol{C}_1)$. Thus $u \notin \rho(\mathcal{T}(\boldsymbol{C}_1)) \setminus \rho(\mathcal{T}(\boldsymbol{C}_2))$, which is a contradiction. Consequently, the proposition must hold true.

Corollary 4.5.1 *In the above,* $B \diamond (\rho(\mathcal{T}(\boldsymbol{C}_1)) \setminus \rho(\mathcal{T}(\boldsymbol{C}_2))) \subseteq A$.

Let $L \subseteq X^*$ be a regular language over X. By $\#L$, we denote the number $min\{|S| \mid \exists \boldsymbol{A} = (S, X, \delta, s_0, F), L = \mathcal{T}(\boldsymbol{A})\}$ where $|S|$ denotes the cardinality of S. Moreover, $\mathcal{I}(n, X)$ denotes the class of languages $\{L \subseteq X^* \mid \#L \leq n\}$.

Theorem 4.5.1 *Let $A \subseteq X^*$ and let n be a positive integer. Then it is decidable whether there exist nontrivial regular languages $B \in \mathcal{I}(n, X)$ and $C \subseteq X^*$ such that $A = B \diamond C$.*

Proof Let $A \subseteq X^*$ be a regular language. Assume that there exist nontrivial regular languages $B \in \mathcal{I}(n, X)$ and $C \subseteq X^*$ such that $A = B \diamond C$. Then by Proposition 4.5.1 and its corollary, $C \subseteq \rho(\mathcal{T}(\boldsymbol{C}_1)) \setminus \rho(\mathcal{T}(\boldsymbol{C}_2))$ and $B \diamond (\rho(\mathcal{T}(\boldsymbol{C}_1)) \setminus \rho(\mathcal{T}(\boldsymbol{C}_2))) \subseteq A$. Hence $A = B \diamond (\rho(\mathcal{T}(\boldsymbol{C}_1)) \setminus \rho(\mathcal{T}(\boldsymbol{C}_2)))$. Thus we have the following algorithm: (1) Choose a nontrivial regular language $B \subseteq X^*$ from $\mathcal{I}(n, X)$ and construct the language $\rho(\mathcal{T}(\boldsymbol{C}_1)) \setminus \rho(\mathcal{T}(\boldsymbol{C}_2))$ (see Lemma 4.5.1). (2)

Let $C = \rho(\mathcal{T}(C_1)) \setminus \rho(\mathcal{T}(C_2))$. (3) Compute $B \diamond C$. (4) If $A = B \diamond C$, then the output is "YES" and "NO", otherwise. (4) If the output is "NO", then choose another element in $\mathcal{I}(n, X)$ as B and continue the procedures (1) - (3). (5) Since $\mathcal{I}(n, X)$ is a finite set, the above process terminates after a finite-step trial. Once one gets the output "YES", then there exist nontrivial regular languages $B \in \mathcal{I}(n, X)$ and $C \subseteq X^*$ such that $A = B \diamond C$. Otherwise, there are no such languages.

From the above theorem, we can obtain the same result by C. Câmpeanu et al. [14] as a corollary.

Corollary 4.5.2 *For a given positive integer n and a regular language $A \subseteq X^*$, the problem whether $A = B \diamond C$ for a nontrivial language $B \in \mathcal{F}(n, X)$ and a nontrivial regular language $C \subseteq X^*$ is decidable.*

Proof Obvious from the fact that $\mathcal{F}(n, X) \subseteq \mathcal{I}(|X|^{n+1}, X)$.

4.6 Shuffle closures

In this section, we will study the shuffle closures of commutative regular languages and present a characterization of those commutative regular languages whose shuffle closures are also regular. Our characterization provides an algorithm to decide whether the shuffle closure of a given commutative regular language is regular. The corresponding decidability problem for general regular languages is still open.

Let $A \subseteq X^*$. By alph(A) we denote the alphabet of A, i.e. $a \in$ alph(A) if and only if a occurs in at least one word of A. We will apply this notation to words as well, i.e. alph(u) denotes the set of those letters which occur in the word u.

Let $N = \{0, 1, \ldots\}$, let $X = \{a_1, \ldots, a_n\}$ and let $u \in X^*$. The mapping Ψ of X^* into the set N^n defined by

$$\Psi(u) = (|u|_{a_1}, \ldots, |u|_{a_n})$$

is called the *Parikh mapping*. For a language $A \subseteq X^*$, we define $\Psi(A) = \{\Psi(u) \mid u \in A\}$. Moreover, if $S \subseteq N^n$, then $\Psi^{-1}(S) = \{u$

$\mid u \in X^*, \Psi(u) \in S\}$. The language A is called *commutative* if $A = \Psi^{-1}\Psi(A)$.

We recall that, by the *shuffle product* of words $u, v \in X^*$, we mean the set $u \diamond v = \{w \mid w = u_1v_1u_2v_2\cdots u_kv_k, u = u_1\cdots u_k, v = v_1\cdots v_k, u_i, v_j \in X^*\}$.

Obviously, $u \diamond v = v \diamond u$ and $(u \diamond v) \diamond w = u \diamond (v \diamond w)$ hold for any $u, v, w \in X^*$. Now, let $A, B \subseteq X^*$. Recall that, by the shuffle product of A and B, we mean the set $\bigcup_{u\in A, v\in B} u\diamond v$ which is denoted by $A\diamond B$. It is easy to see that $A \diamond B = B \diamond A$ and $(A \diamond B) \diamond C = A \diamond (B \diamond C)$ for any $A, B, C \subseteq X^*$. It is worth noting that the shuffle product of commutative languages is also a commutative language. Let $A \subseteq X^*$. The language $A^\diamond$ determined as the union of all finite shuffle products of words in A is called the *shuffle closure* of A. The language obtained by adjoining the empty word to $A^\diamond$ is denoted by A°. It is obvious that if $A = \emptyset$, then $A^\diamond = \emptyset$ and $A^\circ = \{\epsilon\}$. Moreover, if $A = \{\epsilon\}$, then $A^\diamond = A^\circ = \{\epsilon\}$. For the sake of simplicity, if $A = \{u\}$ for some word $u \in X^*$, then we write $u^\diamond$ and u° instead of $\{u\}^\diamond$ and $\{u\}^\circ$, respectively.

We will use the following notations. Let $m \geq 2$ be an arbitrarily fixed integer. We denote the language $A^{(m-1)\diamond} \diamond A$ by $A^{m\diamond}$ where $A^{1\diamond} = A$. Moreover, the language $\bigcup_{1\leq i\leq m} A^{i\diamond}$ is denoted by $A^{m\diamond\downarrow}$. By the definitions, we have $A^\diamond = \bigcup_{i\geq 1} A^{i\diamond}$. For a finite index set $I = \{i_1, i_2, \ldots, i_k\}$ and languages F_{i_t}, $t = 1, \ldots, k$, the language $F_{i_1} \diamond F_{i_2} \diamond \cdots \diamond F_{i_k}$ is denoted by $\prod^\diamond_{j\in I} F_j$. In particular, if $I = \emptyset$, then let $\prod^\diamond_{j\in I} F_j = \{\epsilon\}$.

The following statement has already been proven in Lemma 4.3.1: *If $A, B \subseteq X^*$ are regular languages, then $A \diamond B$ is regular as well.*

To prove our main result, we need some preparation. First of all, observe that if $A = \emptyset$ or $A = \{\epsilon\}$, then both A and $A^\diamond$ are regular. Therefore, in what follows, throughout this section, $A \subseteq X^*$ will denote a language with $A \neq \emptyset$ and $A \neq \{\epsilon\}$.

Lemma 4.6.1 *If $A \subseteq X^*$ is commutative and $A^\diamond$ is regular, then $\epsilon \neq u \in A$, $a \in \mathrm{alph}(u)$ and $a^+ \cap A = \emptyset$ imply $a^tX^* \cap (A \cap \mathrm{alph}(u)^*) \neq \emptyset$ for any positive integer t.*

Proof Let $\epsilon \neq u \in A$, $a \in \mathrm{alph}(u)$, $a^+ \cap A = \emptyset$ and suppose that the statement does not hold. Since $\epsilon \neq u \in A$ and $a \in \mathrm{alph}(u)$, there exists at least one positive integer t such that $a^t X^* \cap (A \cap \mathrm{alph}(u)^*) \neq \emptyset$. On the other hand, observe that $a^t X^* \cap (A \cap \mathrm{alph}(u)^*) \neq \emptyset$ implies $a^j X^* \cap (A \cap \mathrm{alph}(u)^*) \neq \emptyset$ for any $j = 1, \ldots, t$. Thus, from our assumption it follows that there exists a greatest integer $t_0 > 0$ such that $a^{t_0} X^* \cap (A \cap \mathrm{alph}(u)^*) \neq \emptyset$. Consequently, by the commutativity of A, we have $|q|_a \leq t_0$ for any $q \in A \cap \mathrm{alph}(u)^*$.

Now, let $a^{t_0} w \in A$ where $w \in (\mathrm{alph}(u) \setminus \{a\})^*$. By $a^{t_0} X^* \cap (A \cap \mathrm{alph}(u)^*) \neq \emptyset$, such a word exists. Then $a^{ht_0} w^h \in A^\diamond$ for any positive integer h. On the other hand, $A^\diamond$ is regular and thus there exist positive integers k, p with $a^k \equiv a^{k+p}(P_{A^\diamond})$. Choosing an h_0 satisfying $h_0 t_0 \geq k$, by the above relation, we have $a^{h_0 t_0 + mp} w^{h_0} \in A^\diamond$ for any nonnegative integer m. Choose an appropriate large m_0 for which $m_0 p \geq h_0 t_0 |w|$. Since $v_0 = a^{h_0 t_0 + m_0 p} w^{h_0} \in A^\diamond$ and $\mathrm{alph}(v_0) \subseteq \mathrm{alph}(u)$, there exist $u_1, \ldots, u_s \in A \cap \mathrm{alph}(u)^+$ such that

$$v_0 = a^{h_0 t_0 + m_0 p} w^{h_0} \in u_1 \diamond \cdots \diamond u_s.$$

From $a^+ \cap A = \emptyset$ it follows that any word in $A \cap \mathrm{alph}(u)^+$ contains at least one letter from $\mathrm{alph}(u) \setminus \{a\}$. Therefore, by the above relation, we have $s \leq h_0 |w|$. On the other hand, $|q|_a \leq t_0$ for any $q \in A \cap \mathrm{alph}(u)^+$. Consequently, $|u_j|_a \leq t_0$, $j = 1, \ldots, s$ and thus we have $|v|_a \leq st_0 \leq h_0 t_0 |w|$ for any $v \in u_1 \diamond \cdots \diamond u_s$. But $|v_0|_a = h_0 t_0 + m_0 p > h_0 t_0 |w|$ which is a contradiction. This completes the proof of Lemma 4.6.1.

Lemma 4.6.2 *If $A \subseteq X^*$ is a commutative regular language, then there exists a finite nonempty commutative language $F \subseteq A$ such that $A^\diamond = A^{|X|\diamond\downarrow} \diamond F^\circ$.*

Proof Let $X = \{a_1, \ldots, a_n\}$. Without loss of generality, we may assume that $\mathrm{alph}(A) = \{a_1, \ldots, a_r\}$ for some $1 \leq r \leq n$. First, we define a suitable set F. Since A is a regular language, there exist positive integers k, p such that $a_j^k \equiv a_j^{k+p}(P_A)$ for any $j = 1, \ldots, r$. Now, define the set F as follows:

$$F = \{u \mid u \in A, \forall j = 1, \ldots, r, |u|_{a_j} < k + p\}.$$

Then $F \subseteq A$ and F is a finite commutative language. Moreover, it is easy to see that $F \neq \emptyset$. Let $B = A^{n\diamond\downarrow} \diamond F^\circ$. It is obvious that

$A^\diamond \supseteq B$. To prove the converse inclusion, let $u \in A^\diamond$ be an arbitrary word. If $u = \epsilon$, then $u \in B$ holds as well. Assume that $u \neq \epsilon$. Then u can be decomposed into the form $u \in u_1 \diamond \cdots \diamond u_m$ where $m \geq 1$ and $u_j \in A$, $j = 1, \ldots, m$. If $m \leq n$, then we immediately obtain that $u \in B$. Now, assume that $m > n$. Let $|u_i|_{a_j} = t_{ij}$, $i = 1, \ldots, m$, $j = 1, \ldots, r$. Since A and B are commutative, we may assume that $u_i = a_1^{t_{i1}} \cdots a_r^{t_{ir}}$, $i = 1, \ldots, m$. Using the relations $a_j^k \equiv a_j^{k+p}(P_A)$, $j = 1, \ldots, r$, we define a new shuffle decomposition for u as follows.

If $t_{i1} < k + p$ for any $i = 1, \ldots, m$, then let $t'_{i1} = t_{i1}$ for any $i = 1, \ldots, m$. In the opposite case, there exists at least one $i \in \{1, \ldots, m\}$ satisfying $t_{i1} \geq k + p$. Without loss of generality, we may assume that $t_{11} \geq k + p$. Now, for any $i \in \{2, \ldots, m\}$, let $n_i = 0$ if $t_{i1} < k + p$ and let n_i such an integer for which $k \leq t_{i1} - n_i p < k + p$ holds if $t_{i1} \geq k + p$. Moreover, let $t'_{i1} = t_{i1} - n_i p$. Finally, define t'_{11} by $t'_{11} = t_{11} + (n_2 + \cdots + n_m)p$.

Consider the words:

$$u_i^{(1)} = a_1^{t'_{i1}} a_2^{t_{i2}} \cdots a_r^{t_{ir}}, \; i = 1, \ldots, m.$$

Then $u_i^{(1)} \in A$, $i = 1, \ldots, m$ and $u \in u_1^{(1)} \diamond \cdots \diamond u_m^{(1)}$ where $t'_{i1} < k + p$ for any i, $i = 2, \ldots, m$.

Using the same idea as above, we can construct a further sequence of words $u_i^{(2)} = a_1^{s_{i1}} \cdots a_r^{s_{ir}}$, $i = 1, \ldots, m$, such that $u_i^{(2)} \in A$, $i = 1, \ldots, m$ and $u \in u_1^{(2)} \diamond \cdots \diamond u_m^{(2)}$ where $s_{i1} < k + p$ if $i \in \{2, \ldots, m\}$ and $s_{i2} < k + p$ if $i \in \{3, \ldots, m\}$. Continuing this procedure, finally we obtain such a sequence of words $u_i^{(r)} = a_1^{p_{i1}} \cdots a_r^{p_{ir}}$, $i = 1, \ldots, m$, for which $u_i^{(r)} \in A$, $i = 1, \ldots, m$ with $u \in u_1^{(r)} \diamond \cdots \diamond u_m^{(r)}$ and, for any $t \in \{1, \ldots, r\}$, $p_{it} < k + p$ holds if $i \in \{1 + t, \ldots, m\}$. Now, observe that $u_{r+1}^{(r)}, \ldots, u_m^{(r)} \in F$ and thus we have $u \in A^{r\diamond} \diamond F^\diamond \subseteq A^{n\diamond\downarrow} \diamond F^\diamond$. Consequently, $u \in B$ which completes the proof of the lemma.

Lemma 4.6.3 *Let $A \subseteq X^*$ be a commutative regular language and $Z = \{a \mid a \in X, a^+ \cap A \neq \emptyset\}$. If $\epsilon \neq u \in A$, $a \in \mathrm{alph}(u)$ and $a^+ \cap A = \emptyset$ imply $a^t X^* \cap (A \cap \mathrm{alph}(u)^*) \neq \emptyset$ for any positive integer t, then there exists a positive integer M such that*

$$w^\diamond \subseteq A^{M\diamond\downarrow} \diamond \prod_{a \in Z} {}^\diamond(a^* \cap A^*)$$

holds for any $w \in A$.

Proof Denote the elements of Z by $b_1, \ldots, b_s$ and let $Y = X \setminus Z = \{a_1, \ldots, a_r\}$. We notice that $Z \cup Y \neq \emptyset$. By the regularity of A, there exist positive integers k, p' such that $a_j^k \equiv a_j^{k+p'}(P_A)$ for any $j = 1, \ldots, r$ and $b_i^k \equiv b_i^{k+p'}(P_A)$ for any i, $i = 1, \ldots, s$. On the other hand, for any $b_i \in \{b_1, \ldots, b_s\}$, we have $b_i^+ \cap A \neq \emptyset$ and thus there exists a smallest positive integer p_i with $b_i^{p_i} \in A$. Let $t = \text{l.c.m.}\{p_1, \ldots, p_s\}$ if $Z \neq \emptyset$. Then $b_i^t \in A^*$ for any $i = 1, \ldots, s$. Now, let

$$p = \begin{cases} p' & \text{if } Z = \emptyset, \\ \text{l.c.m.}\{p', t\} & \text{otherwise.} \end{cases}$$

Then we have

$$\begin{array}{ll} & a_j^k \equiv a_j^{k+p}(P_A),\ j = 1, \ldots, r, \\ (1) & b_i^k \equiv b_i^{k+p}(P_A),\ i = 1, \ldots, s, \\ & b_i^p \in A^*,\ i = 1, \ldots, s. \end{array}$$

Let f be the smallest positive integer for which $fp \geq (k+p)n$ holds where $n = |X|$. We define the integers q and M by $q = k + fp$ and $M = np + pq - 1$.

Now, we are ready to prove the lemma. For this purpose, let $w \in A$ be an arbitrary word. If $w = \epsilon$, then obviously the lemma holds true. Assume that $w \neq \epsilon$. Depending on $\text{alph}(w)$, we distinguish the following three cases.

Case 1: $\text{alph}(w) \cap Y \neq \emptyset$ and $\text{alph}(w) \cap Z \neq \emptyset$. Without loss of generality, we may assume that $\text{alph}(w) \cap Y = \{a_1, \ldots, a_d\}$ and $\text{alph}(w) \cap Z = \{b_1, \ldots, b_e\}$ for some positive integers $d \leq r$ and $e \leq s$. By our assumption, there exist $x_1, \ldots, x_d \in \text{alph}(w)^*$ such that $a_j^q x_j \in A$, $j = 1, \ldots, d$. Moreover, by (1), we may assume that, for any $g = 1, \ldots, d$, $|x_g|_{a_j} < k + p$, $j = 1, \ldots, d$. Let $j \in \{1, \ldots, d\}$ be an arbitrary integer. From (1) and the definition of q, it follows that A contains the members of the following sequence:

$$a_j^{q-fp}x_j,\ a_j^{q-(f-1)p}x_j,\ a_j^{q-(f-2)p}x_j,\ a_j^{q-(f-3)p}x_j, \cdots.$$

Now, observe that, for any m_j, $m_j \in \{-fp, -fp+1, -fp+2, \cdots\}$, there are not necessarily distinct p words in the above sequence such that their shuffle product contains the word $a_j^{pq+m_jp}x_j^p$. Therefore,

$$a_j^{pq+m_jp}x_j^p \in A^{p\diamond},\ m_j = -fp, -fp+1, -fp+2, -fp+3, \cdots.$$

Consider the above sequence of words for any $j = 1, \ldots, d$. Now, taking any shuffle product which contains one and only one member from each sequence, we have

$$a_1^{pq+m_1p}a_2^{pq+m_2p}\cdots a_d^{pq+m_dp}x_1^p\cdots x_d^p \in A^{dp\diamond}$$

for any $(m_1, \ldots, m_d) \in \{-fp, -fp+1, -fp+2, -fp+3, \cdots\}^d$. Since $|x_g|_{a_j} < k+p$, $j = 1, \ldots, d$ hold for any $g = 1, \cdots, d$,

$$a_1^{t_1p}a_2^{t_2p}\cdots a_d^{t_dp}b_1^{v_1p}\cdots b_e^{v_ep} \in \Psi^{-1}\Psi(x_1^p\cdots x_d^p)$$

where $t_j < (k+p)d$, $j = 1, \ldots, d$ and $v_i < (k+p)d$, $i = 1, \ldots, e$. Since $A^{dp\diamond}$ is commutative,

$$a_1^{pq+(m_1+t_1)p}a_2^{pq+(m_2+t_2)p}\cdots a_d^{pq+(m_d+t_d)p}b_1^{v_1p}\cdots b_e^{v_ep} \in A^{dp\diamond}$$

for any $(m_1, \ldots, m_d) \in \{-fp, -fp+1, -fp+2, \cdots\}^d$. Then by $t_j < (k+p)d$, $j = 1, \ldots, d$ and the definition of f, for arbitrary positive integers $\delta_1, \ldots, \delta_d$, there exist $m_j \in \{-fp, -fp+1, -fp+2, -fp+3, \cdots\}$, $j = 1, \ldots, d$, such that $pq + (m_j + t_j)p = \delta_j pq$, $j = 1, \ldots, d$. Consequently,

$$a_1^{\delta_1pq}\cdots a_d^{\delta_dpq}b_1^{v_1p}\cdots b_e^{v_ep} \in A^{dp\diamond} \tag{2}$$

for any $(\delta_1, \ldots, \delta_d) \in \{1, 2, \cdots\}^d$.

Now, let m be an arbitrary positive integer. We will prove that

$$w^{mpq\diamond} \subseteq A^{np\diamond\downarrow} \diamond \prod_{a\in Z} {}^{\diamond}(a^* \cap A^*).$$

By our assumption on w, $|w|_{a_j} = \alpha_j$, $j = 1, \ldots, d$, and $|w|_{b_i} = \beta_i$, $i = 1, \ldots, e$ for some positive integers $\alpha_1, \ldots, \alpha_d, \beta_1, \ldots, \beta_e$. Let $u = a_1^{\alpha_1}\cdots a_d^{\alpha_d}b_1^{\beta_1}\cdots b_e^{\beta_e}$. Then $\Psi^{-1}\Psi(w) = \Psi^{-1}\Psi(u)$. Let $u_0 = a_1^{\alpha_1mpq}\cdots a_d^{\alpha_dmpq}b_1^{\beta_1mpq}\cdots b_e^{\beta_empq}$. Then $w^{mpq\diamond} \subseteq \Psi^{-1}\Psi(u_0)$. By the definitions of f and q, we have $\beta_i mq \geq k+fp \geq (k+p)n \geq (k+p)d > v_i$, $i = 1, \ldots, d$. Therefore,

$$u_1 = a_1^{\alpha_1 mpq} \cdots a_d^{\alpha_d mpq} b_1^{v_1 p} \cdots b_e^{v_e p} b_1^{\gamma_1 p} \cdots b_e^{\gamma_e p} \in \Psi^{-1}\Psi(u_0)$$

where $\gamma_i = \beta_i mq - v_i > 0$, $i = 1, \ldots, e$. Now, observe that, by (2), we have

$$a_1^{\alpha_1 mpq} \cdots a_d^{\alpha_d mpq} b_1^{v_1 p} \cdots b_e^{v_e p} \in A^{dp\diamond}.$$

On the other hand, by the definition of p,

$$b_1^{\gamma_1 p} \cdots b_e^{\gamma_e p} \in \prod_{a \in Z} {}^{\diamond}(a^* \cap A^*).$$

Therefore, $u_1 \in A^{dp\diamond} \diamond \prod_{a\in Z}^{\diamond}(a^* \cap A^*)$. By the commutativity of $A^{dp\diamond} \diamond \prod_{a\in Z}^{\diamond}(a^* \cap A^*)$, we have

$$\Psi^{-1}\Psi(u_1) \subseteq A^{dp\diamond} \diamond \prod_{a \in Z} {}^{\diamond}(a^* \cap A^*).$$

But $w^{mpq\diamond} \subseteq \Psi^{-1}\Psi(u_0) = \Psi^{-1}\Psi(u_1)$ and $d \leq n$ which imply the required inclusion, i.e.

(3) $$w^{mpq\diamond} \subseteq A^{np\diamond\downarrow} \diamond \prod_{a \in Z} {}^{\diamond}(a^* \cap A^*)$$

holds, for an arbitrary positive integer m.

Since $w^{\diamond} = \bigcup_{i \geq 1} w^{i\diamond}$, it is enough to prove that $w^{i\diamond} \subseteq A^{M\diamond\downarrow} \diamond \prod_{a\in Z}^{\diamond}(a^* \cap A^*)$ for any i, $i = 1, 2, \cdots$. For this purpose, let i be an arbitrarily fixed positive integer. If $i < pq$, then

$$w^{i\diamond} \subseteq A^{i\diamond} \subseteq A^{(pq-1)\diamond\downarrow} \subseteq A^{M\diamond\downarrow} \diamond \prod_{a \in Z} {}^{\diamond}(a^* \cap A^*).$$

If $i \geq pq$, then there exist integers $m > 0$ and $0 \leq \alpha < pq$ such that $i = mpq + \alpha$. On the other hand, $w^{i\diamond} = w^{mpq\diamond} \diamond w^{\alpha\diamond}$. Now, $w^{\alpha\diamond} \subseteq A^{(pq-1)\diamond\downarrow}$. Moreover, $w^{mpq\diamond} \subseteq A^{np\diamond\downarrow} \diamond \prod_{a\in Z}^{\diamond}(a^* \cap A^*)$ from (3). Therefore,

$$w^{mpq\diamond} \diamond w^{\alpha\diamond} \subseteq A^{np\diamond\downarrow} \diamond \prod_{a \in Z} {}^{\diamond}(a^* \cap A^*) \diamond A^{(pq-1)\diamond\downarrow} \subseteq A^{M\diamond\downarrow} \diamond \prod_{a \in Z} {}^{\diamond}(a^* \cap A^*)$$

which completes the proof of Case 1.

Case 2: $\mathrm{alph}(w) \cap Z = \emptyset$. Without loss of generality, we may assume that $\mathrm{alph}(w) \cap Y = \{a_1, \ldots, a_d\}$. Observe that $d > 1$. Indeed, in the opposite case, $a_1^t x_1 \in A$ for some $x_1 \in \mathrm{alph}(w)^* = a_1^*$ and $t > 0$ from our assumption and thus $a_1^{t+|x_1|} \in A$ which contradicts the fact that $a_1 \in Y$. Then in a similar way as Case 1, we have

$$(4) \qquad a_1^{\delta_1 pq} \cdots a_d^{\delta_d pq} \in A^{dp\diamond}$$

for any $(\delta_1, \ldots, \delta_d) \in \{1, 2, \cdots\}^d$. In this case, $w \in \Psi^{-1}\Psi(a_1^{\alpha_1} \cdots a_d^{\alpha_d})$ where $\alpha_j = |w|_{a_j} > 0$, $j = 1, \ldots, d$. Now, let m be a positive integer. Then $w^{mpq\diamond} \subseteq \Psi^{-1}\Psi(a_1^{\alpha_1 mpq} \cdots a_d^{\alpha_d mpq})$. On the other hand, $a_1^{\alpha_1 mpq} \cdots a_d^{\alpha_d mpq} \in A^{dp\diamond}$ from (4). Since $A^{dp\diamond}$ is commutative, we have $w^{mpq\diamond} \subseteq A^{dp\diamond} \subseteq A^{np\diamond\downarrow}$ for any positive integer m. Now, in the same way as above, it can be proved that $w^{i\diamond} \subseteq A^{M\diamond\downarrow}$ holds for any i, $i = 1, 2, \cdots$ and thus

$$w^{\diamond} \subseteq A^{M\diamond\downarrow} \subseteq A^{M\diamond\downarrow} \diamond \prod_{a\in Z}{}^{\diamond}(a^* \cap A^*)$$

which completes the proof of this case.

Case 3: $\mathrm{alph}(w) \cap Y = \emptyset$. Without loss of generality, we may assume that $\mathrm{alph}(w) \cap Z = \{b_1, \ldots, b_e\}$. Then $w \in \Psi^{-1}\Psi(b_1^{\beta_1} \cdots b_e^{\beta_e})$ where $\beta_i = |w|_{b_i} > 0$, $i = 1, \ldots, e$. Now, let m be an arbitrary positive integer. We show that

$$w^{mpq\diamond} \subseteq A^{M\diamond\downarrow} \diamond \prod_{a\in Z}{}^{\diamond}(a^* \cap A^*).$$

If $m = 1$, then $w^{pq\diamond} \subseteq A^{pq\diamond} \subseteq A^{M\diamond\downarrow} \diamond \prod_{a\in Z}^{\diamond}(a^* \cap A^*)$. Now, let $m > 1$. Then

$$w^{mpq\diamond} \subseteq \Psi^{-1}\Psi(b_1^{\beta_1 pq} \cdots b_e^{\beta_e pq}) \diamond \Psi^{-1}\Psi(b_1^{\beta_1(m-1)pq} \cdots b_e^{\beta_e(m-1)pq}).$$

Observe that $\Psi^{-1}\Psi(b_1^{\beta_1 pq} \cdots b_e^{\beta_e pq}) \subseteq A^{pq\diamond} \subseteq A^{M\diamond\downarrow}$. Furthermore, by the definition of p, we have $b_i^{\beta_i(m-1)pq} \in A^*$, $i = 1, \ldots, e$ and thus

$$\Psi^{-1}\Psi(b_1^{\beta_1(m-1)pq}\cdots b_e^{\beta_e(m-1)pq}) \subseteq \prod_{a\in Z}{}^{\diamond}(a^*\cap A^*).$$

Consequently,

$$w^{mpq\diamond} \subseteq A^{M\diamond\downarrow}\diamond\prod_{a\in Z}{}^{\diamond}(a^*\cap A^*)$$

holds for any positive integer m. Using this inclusion, we can prove, in the same way as the previous cases, that $w^{i\diamond}\subseteq A^{M\diamond\downarrow}\diamond\prod^{\diamond}_{a\in Z}(a^*\cap A^*)$ for any i, $i=1,2,\cdots$, which implies

$$w^{\diamond} \subseteq A^{M\diamond\downarrow}\diamond\prod_{a\in Z}{}^{\diamond}(a^*\cap A^*).$$

This completes the proof of Lemma 4.6.3.

Now, we are ready to prove the following theorem.

Theorem 4.6.1 *Let $A\subseteq X^*$ be a commutative regular language. Then $A^{\diamond}$ is regular if and only if $\epsilon\neq u\in A$, $a\in \mathrm{alph}(u)$ and $a^+\cap A=\emptyset$ imply $a^tX^*\cap(A\cap \mathrm{alph}(u)^*)\neq\emptyset$ for any positive integer t.*

Proof (⇒) Obvious from Lemma 4.6.1.

(⇐) Assume that the above condition holds. We prove that $A^{\diamond}$ is regular.

Let $X=\{a_1,\ldots,a_n\}$. Then by Lemma 4.6.2, we have $A^{\diamond}=A^{n\diamond\downarrow}\diamond F^{\diamond}$ where $F\subseteq A$ and F is a finite nonempty commutative language. It is easy to see that $F^{\diamond}=\prod^{\diamond}_{u\in F}u^{\diamond}$. By Lemma 4.6.3, there exists an $M>0$ such that $u^{\diamond}\subseteq A^{M\diamond\downarrow}\diamond\prod^{\diamond}_{a\in Z}(a^*\cap A^*)$ for any $u\in F$. Consequently, $F^{\diamond}\subseteq\{\epsilon\}\cup A^{M|F|\diamond\downarrow}\diamond\prod^{\diamond}_{a\in Z}(a^*\cap A^*)$, and thus we have

$$A^{\diamond}\subseteq A^{(n+M|F|)\diamond\downarrow}\diamond\prod_{a\in Z}{}^{\diamond}(a^*\cap A^*).$$

On the other hand, each member of the right-side is contained in $A^{\diamond}$, and hence

$$A^{\diamond} = A^{(n+M|F|)\diamond\downarrow} \diamond \prod_{a\in Z}{}^{\diamond}(a^* \cap A^*).$$

Finally, observe that, by Lemma 4.6.2, both $A^{(n+M|F|)\diamond\downarrow}$ and $\prod^{\diamond}_{a\in Z}(a^* \cap A^*)$ are regular and thus $A^{\diamond}$ is regular as well.

This completes the proof of Theorem 4.6.1.

Using Lemma 4.6.1 and the same idea as in the above proof, we obtain the following theorem which presents the fact that if $A^{\diamond}$ is regular, then it is equal to a shuffle product of a union of finite shuffle powers of A and a regular language. More precisely, we have the following result.

Theorem 4.6.2 *Let $A \subseteq X^*$ be a commutative regular language. Moreover, let $Z = \{a \mid a \in \mathrm{alph}(A), a^+ \cap A \neq \emptyset\}$. Then $A^{\diamond}$ is regular if and only if there exists a positive integer N such that $A^{\diamond} = A^{N\diamond\downarrow} \diamond \prod^{\diamond}_{a\in Z}(a^* \cap A^*)$.*

Notice that it is decidable whether or not the condition of Theorem 4.6.1 is satisfied and thus it is decidable whether $A^{\diamond}$ is regular for a commutative regular language A. Using Theorem 4.6.2, we can provide another algorithm to decide whether $A^{\diamond}$ is regular for a commutative regular language $A \subseteq X^*$ as follows: In Theorem 4.6.2, the value of N is given by A, e.g. $N = (n+1)p(k+p+1)(k+p)^p$. Using this N, we can check whether or not $A^{(N+1)\diamond\downarrow} \subseteq A^{N\diamond\downarrow} \diamond \prod^{\diamond}_{a\in Z}(a^* \cap A^*)$. If $A^{(N+1)\diamond\downarrow} \subseteq A^{N\diamond\downarrow} \diamond \prod^{\diamond}_{a\in Z}(a^* \cap A^*)$, then $A^{\diamond}$ is regular. Otherwise, $A^{\diamond}$ is not regular.

Chapter 5

Shuffle Closed Languages

In the previous chapter, we dealt with the shuffle closure of a commutative regular language. The shuffle closure of a language is a *shuffle closed* language defined below. In this chapter, we will deal with the structures of commutative regular shuffle-closed and regular strongly shuffle-closed languages. Moreover, we will prove that every hypercode is finite.

The contents of this chapter are based on the results in [47].

5.1 Sh-closed and ssh-closed languages

Let $L \subseteq X^*$ and let $u, v \in X^*$. We consider the following conditions on a nonempty language L:

$$(1)\ u, v \in L \Rightarrow u \diamond v \subseteq L. \ (2)\ u \in L, u \diamond v \cap L \neq \emptyset \Rightarrow v \in L.$$

Definition 5.1.1 A language satisfying the condition (1) is said to be *shuffle closed* or *sh-closed.* A language satisfying the conditions (1) and (2) is said to be *strongly shuffle closed* or *ssh-closed.*

Every sh-closed language is a subsemigroup of X^* and, since $u \diamond \epsilon = u$, every ssh-closed language contains ϵ and is consequently a submonoid of X^*.

Example 5.1.1 Let $X = \{a, b\}$ and let $k \geq 1$. Then: (1) $L_k(a) = \{u \in X^+ \mid |u|_a \geq k\}$ is sh-closed, but not ssh-closed if $k \geq 2$. (2) $L_{ab} = \{u \in X^+ \mid |u|_a = |u|_b\}$ is ssh-closed.

Let $L \subseteq X^*$. Then $L^\diamond$ and L° are sh-closed. Moreover, $L^\diamond$ is the smallest sh-closed language containing L.

Fact *(1)* $L_1 \subseteq L_2 \Rightarrow L_1^\diamond \subseteq L_2^\diamond$ *and* $L_1^\circ \subseteq L_2^\circ$ *for any* $L_1, L_2 \subseteq X^*$. *(2)* $(L^\diamond)^\diamond = L^\diamond$ *and* $(L^\circ)^\circ = L^\circ$ *for any* $L \subseteq X^*$.

If $L \subseteq X^*$, then the following properties are equivalent:
(1) $xuvy \in L$ *implies* $xvuy \in L$. *(2) If* $u \in L$, *then* $\sigma(u) \subseteq L$ *where* $\sigma(u)$ *is the set of all permutations of the letters of* u.

A language having one of the above equivalent properties is called a *commutative* language.

Proposition 5.1.1 *Every ssh-closed language is commutative.*

Proof Let $xuvy \in L$. Since L is sh-closed, $xxuvuvyy \in L$ and since L is ssh-closed, $xvuy \in L$.

Every submonoid of X^* that is a commutative language is sh-closed, but it is not in general ssh-closed. For example, the language $L = \{aaa, ab, ba\}^\circ$ is sh-closed and commutative, but not ssh-closed.

5.2 Sh-free languages and sh-bases

A language $L \subseteq X^+$ is called *sh-free* if it is nonempty and if $u_1, u_2, \ldots, u_k \in L$, $k \geq 2$, implies $u_1 \diamond u_2 \diamond \cdots \diamond u_k \cap L = \emptyset$.

A typical class of sh-free languages is the class of hypercodes. A *hypercode* H is a nonempty subset of X^+ such that $u_1u_2\cdots u_k \in H$, $x_1u_1x_2u_2\cdots u_kx_{k+1} \in H$ implies $x_1 = x_2 = \cdots = x_{k+1} = \epsilon$ (see [64] - [66]).

In Chapter 6 (Section 6.2), we will introduce the notion of codes. A hypercode is a code and it is sh-free. An sh-free language is not necessarily a code. For example, if $X = \{a, b, c, d\}$, then $A = \{ab, abc, cd, d\}$ is sh-free but it is not a code.

Definition 5.2.1 Let $L \subseteq X^*$ be an sh-closed language. Then by $J(L)$ we denote the language $(L \setminus \{\epsilon\}) \setminus ((L \setminus \{\epsilon\}) \diamond (L \setminus \{\epsilon\}))$ and we call it the *shuffle base* (or in short, *sh-base*) of L.

It is obvious that the sh-base $J(L)$ of an sh-closed language is sh-free. The following result shows that, if L is a regular sh-closed language, then its shuffle base is also regular.

Proposition 5.2.1 *If L is a regular sh-closed language, then its sh-base $J(L)$ is regular and $L = J(L)^\diamond$ or $L = J(L)^\circ$.*

Proof Since L is regular, $L \setminus \{\epsilon\}$ and $(L \setminus \{\epsilon\}) \diamond (L \setminus \{\epsilon\})$ are regular and thus $J(L)$ is regular. From the definition of sh-base, it follows that $L = J(L)^\diamond$ if $\epsilon \notin L$ and $L = J(L)^\circ$ if $\epsilon \in L$.

Lemma 5.2.1 *Let $x \in u \diamond v$ and let $x = x_1 x_2 \cdots x_p$ where $x_i \in X^*$, $i = 1, 2, \ldots, p$. Then there exist $u_1, u_2, \ldots, u_p, v_1, v_2, \ldots, v_p \in X^*$ such that $u = u_1 u_2 \cdots u_p$, $v = v_1 v_2 \cdots v_p$, $x_1 \in u_1 \diamond v_1, x_2 \in u_2 \diamond v_2, \ldots, x_p \in u_p \diamond v_p$.*

Proof We prove the lemma by induction and for that it is enough to deal with the case $p = 2$. Let $x \in u \diamond v$. Then it is easy to see that $x = u_1 v_1 u_2 v_2 \cdots u_r v_r$ where $u_i, v_i \in X \cup \{\epsilon\}$, $i = 1, 2, \ldots, r$. $u = u_1 u_2 \cdots u_r$ and $v = v_1 v_2 \cdots v_r$. Let $x = yz$ where $y, z \in X^*$. Then there exists some $i = 1, 2, \ldots, k$ such that $y = u_1 v_1 u_2 v_2 \cdots u_{k-1} v_{k-1} u_k$ or $y = u_1 v_1 u_2 v_2 \cdots u_{k-1} v_{k-1} u_k v_k$. Let $y = u_1 v_1 u_2 v_2 \cdots u_{k-1} v_{k-1} u_k$. Then let $u_1' = u_1 u_2 \cdots u_{k-1} u_k$, $u_2' = u_{k+1} \cdots u_r$, $v_1' = v_1 v_2 \cdots v_{k-1}$ and let $v_2' = v_k v_{k+1} \cdots v_r$. Obviously, $u = u_1' u_2'$, $v = v_1' v_2'$, $y \in u_1' \diamond v_1'$ and $z \in u_2' \diamond v_2'$. Now if $y = u_1 v_1 u_2 v_2 \cdots u_{k-1} v_{k-1} u_k v_k$, then let $u_1' = u_1 u_2 \cdots u_k$, $u_2' = u_{k+1} \cdots u_r$, $v_1' = v_1 v_2 \cdots v_k$ and let $v_2' = v_{k+1} \cdots v_r$. Obviously, $u = u_1' u_2'$, $v = v_1' v_2'$, $y \in u_1' \diamond v_1'$ and $z \in u_2' \diamond v_2'$. This completes the proof of the lemma.

Lemma 5.2.2 *Let $x \in y_1 \diamond y_2 \diamond \cdots \diamond y_k$ for some $x, y_1, y_2, \ldots, y_k \in X^*$. If $x \sim x'$, i.e. if $x = x_1 x_2 x_3 x_4$ and $x' = x_1 x_3 x_2 x_4$ for some $x_1, x_2, x_3, x_4 \in X^*$, then there exist $y_1', y_2', \ldots, y_k' \in X^*$ such that $y_1 \sim y_1', y_2 \sim y_2', \ldots, y_k \sim y_k'$ and $x' \in y_1' \diamond y_2' \diamond \cdots \diamond y_k'$.*

Proof We prove the lemma by induction and for that it is enough to deal with the case $k = 2$. Let $x \in y \diamond z$ for some $x, y, z \in X^*$. Assume $x \sim x'$. Since $x \in y \diamond z$, by Lemma 5.2.1 there exist $y_1, y_2, y_3, y_4, z_1, z_2, z_3,$

$z_4 \in X^*$ such that $y = y_1y_2y_3y_4$, $z = z_1z_2z_3z_4$, $x_1 \in y_1 \diamond z_1$, $x_2 \in y_2 \diamond z_2$, $x_3 \in y_3 \diamond z_3$ and $x_4 \in y_4 \diamond z_4$. Therefore, $x' = x_1x_3x_2x_4 \in (y_1 \diamond z_1)(y_3 \diamond z_3)(y_2 \diamond z_2)(y_4 \diamond z_4) \subseteq y_1y_3y_2y_4 \diamond z_1z_3z_2z_4$. Notice $y \sim y_1y_3y_2y_4$ and $z \sim z_1z_3z_2z_4$. This completes the proof of the lemma.

Proposition 5.2.2 *If $L \subseteq X^*$ is sh-closed, then L is commutative if and only if the sh-base $J(L)$ of L is commutative.*

Proof ($\Rightarrow$) Let $xuvy \in J(L)$. Since L is commutative, $xvuy \in L$. Suppose $xvuy \notin J(L)$. Then $xvuy \in p \diamond q$ for some $p, q \in L \cap X^+$. By Lemma 5.2.2, $xuvy \in p' \diamond q'$ where $p \sim p'$ and $q \sim q'$. From the assumption that L is commutative, we have $p', q' \in L \cap X^+$. This contradicts the assumption that $xuvy \in J(L)$. Hence $J(L)$ must be commutative.

($\Leftarrow$) Let $xuvy \in L$. Then $xuvy \in y_1 \diamond y_2 \diamond \cdots \diamond y_k$ for some $y_1, y_2, \ldots y_k \in J(L)$. By Lemma 5.2.2, we have $xvuy \in y_1' \diamond y_2' \diamond \cdots \diamond y_k'$ where $y_i \sim y_i', i = 1, 2, \ldots k$. Since $J(L)$ is commutative, $y_i' \in J(L)$ and hence $xvuy \in J(L)^\diamond \subseteq L$. Therefore, L is commutative.

Proposition 5.2.3 *If a language $L \neq \{\epsilon\}$ is ssh-closed, then the sh-base $J(L)$ of L is a commutative hypercode.*

Proof By Proposition 5.1.1, L is commutative. Moreover, from Proposition 5.2.2, it follows that $J(L)$ is commutative. Let $u, v \in J(L)$ where $u \neq v$. Suppose that $u \leq_h v$, i.e. $v \in u \diamond w$ for some $w \in X^*$. Then, since L is ssh-closed, $w \in L$. This contradicts the assumption that $v \in J(L)$. Therefore, $J(L)$ is a hypercode.

The sh-base of the language L_{ab} is the commutative hypercode $\{ab, ba\}$ and $L_{ab} = \{ab, ba\}^\diamond$.

Fact *The sh-closure of a commutative hypercode is not always an ssh-closed language.*

Example 5.2.1 Let $X = \{a, b\}$ and let $H = X^2 \backslash a^2$. Then H is a commutative hypercode, but $H^\diamond$ is not ssh-closed. Notice that $b^2 \in H$ and $abab \in a^2 \diamond b^2$. Moreover, $abab \in ab \diamond ab \subseteq H^\diamond$. Therefore, if $H^\diamond$ is ssh-closed, then $a^2 \in H^\diamond$, which is a contradiction.

Recall that $alph(L)$ denotes the set of all elements of X occurring in words of L.

Proposition 5.2.4 *Let $L \subseteq X^*$ be regular and sh-closed. If $J(L)$, the sh-base of L, is finite, then $a^+ \cap J(L) \neq \emptyset$ for any $a \in alph(L)$.*

Proof Since L is regular, for any $a \in alph(L)$ there exist $k, p \geq 1$ such that $a^k \equiv a^{k+p}$ (P_L) where P_L is the principal congruence defined by L. From $a \in alph(L)$, it follows that $xay \in J(L)$ for some $x, y \in alph(L)^*$. Then $x^k a^k y^k \in L$. Hence $x^k a^{k+np} y^k \in L$ for any $n \geq 1$. Let $\mu = \max\{|x| \mid x \in J(L)\}$ and $k + np > (\mu - 1)k\,|xy|$. Then it is easy to see that $x^k a^{k+np} y^k \in x_1 \diamond x_2 \diamond \cdots \diamond x_s$, $x_i \in J(L)$, $i = 1, 2, \ldots, s$ and $x_j \in a^+$ for some $j = 1, 2, \ldots, s$. This means that $a^+ \cap J(L) \neq \emptyset$.

Fact *The above proposition does not hold true if $J(L)$ is infinite. Let $X = \{a, b\}$ and let $L = (a \cup b^+a)^\circ$. Then $J(L) = a \cup b^+a$ and $L = (a \cup b^+a)^*$. Hence L is regular, but $b^+ \cap J(L) = \emptyset$.*

5.3 Commutative regular languages

In this section, we will determine the structure of commutative regular sh-closed languages.

Let X be a finite alphabet with $|X| = r$ and let $\Psi : X^* \to N^r$ be the Parikh mapping of X^* onto N^r.

The following proposition is immediate.

Proposition 5.3.1 *(1) A language L is commutative if and only if $\Psi^{-1}\Psi(L) = L$. (2) A language L is sh-closed and commutative if and only if $\Psi(L)$ is a subsemigroup of N^r and $\Psi^{-1}\Psi(L) = L$. (3) A nonempty language $H \subseteq X^+$ is a commutative hypercode if and only if $\Psi^{-1}\Psi(H) = H$ and any two distinct elements of $\Psi(H)$ are incomparable with respect to the following partial order on N^r: $(m_1, m_2, \ldots, m_r) \leq (n_1, n_2, \ldots, n_r) \Leftrightarrow m_1 \leq n_1, m_2 \leq n_2, \ldots, m_r \leq n_r$.*

Proposition 5.3.2 *Every commutative code $C \subseteq X^*$ is a hypercode.*

Proof Suppose C is not a hypercode. Then $x \leq_h y$ for some $x, y \in C$, $x \neq y$ where $\leq_h$ is the embedding order. Let $\Psi(x) = (n_1, n_2, \ldots, n_r)$ and let $\Psi(y) = (m_1, m_2, \ldots, m_r)$. Since $x \leq_h y$, $n_1 \leq m_1, n_2 \leq m_2, \ldots, n_r \leq m_r$. Let $x' \in \Psi^{-1}(m_1 - n_1, m_2 - n_2, \ldots, m_r - n_r)$. Then $\Psi(xx') = \Psi(x'x) = (m_1, m_2, \ldots, m_r) = \Psi(y)$. Since C is commutative, $xx', x'x \in C$. Now consider $xx'x \in C^+$. From the equality $(xx')x = x(x'x)$, it follows that C is not a code, which is a contradiction. Therefore, C must be a hypercode.

Proposition 5.3.3 *Let $C \subseteq X^+$ be a code, let $L = C^\circ$ and let $J(L)$ be the sh-base of L. Then L is commutative if and only if $J(L)$ is a commutative hypercode.*

Proof ($\Leftarrow$) Obvious.

($\Rightarrow$) It is obvious that H is a commutative code. By Proposition 5.3.2, $J(L)$ is a hypercode.

Every submonoid (subsemigroup) M of X^* has a unique minimal generating set K, called the *root* of M, and $M = K^*$ ($M = K^+$).

Remark 5.3.1 Let $L \subseteq X^*$ be an sh-closed language. Then L is a subsemigroup of X^*. Let K be the root of L. Since $K = (L^2 \setminus \{\epsilon\}) \setminus (L \setminus \{\epsilon\})$, L is regular if and only if K is regular.

Recall that the shuffle product of two regular languages is regular.

Lemma 5.3.1 *Let $\{a_1, a_2, \ldots, a_r\} \subseteq X$, let $p_1, p_2, \ldots, p_r \geq 0$ and let $x \in X^*$. Then $\Psi^{-1}\Psi(x(a_1^{p_1})^*(a_2^{p_2})^* \cdots (a_r^{p_r})^*) = \bigcup_{y \in \Psi^{-1}\Psi(x)} ((\cdots((y \diamond (a_1^{p_1})^*) \diamond (a_2^{p_2})^*) \cdots) \diamond (a_r^{p_r})^*$.*

Proof Obvious from the above remark.

Lemma 5.3.2 *Let $\{a_1, a_2, \ldots, a_r\} \subseteq X$, let $p_1, p_2, \ldots, p_r \geq 0$ and let $x \in X^*$. Then $\Psi^{-1}\Psi\left(x(a_1^{p_1})^*(a_2^{p_2})^* \cdots (a_r^{p_r})^*\right)$ is regular.*

Proof Obvious from Lemma 5.3.1.

Definition 5.3.1 Let $\alpha, \beta \in N^r$ and let $p_1, p_2, \ldots, p_r \geq 0$. Here $\alpha \stackrel{\rightarrow}{\equiv} \beta \pmod{(p_1, p_2, \ldots, p_r)}$ means that $a_i \geq b_i$ and $a_i \equiv b_i \pmod{p_i}$ for any $i = 1, 2, \ldots, r$ where $\alpha = (a_1, a_2, \ldots, a_r)$ and $\beta = (b_1, b_2, \ldots, b_r)$.

We determine now the structure of a commutative regular sh-closed language $L \subseteq X^*$. Without loss of generality, we can assume that $\epsilon \in L$. Let $\alpha = (m_1, m_2, \ldots, m_r)$ and let $\beta = (n_1, n_2, \ldots, n_r)$ where $m_i, n_j, i, j = 1, 2, \ldots, r$ are nonnegative integers. By $\alpha \leq \beta$ we mean that $m_i \leq n_i$ for any $i = 1, 2, \ldots, r$.

Proposition 5.3.4 *Let $L \subseteq X^*$ with $\epsilon \in L$. Then L is a commutative regular sh-closed language if and only if L is represented as*

$$L = \bigcup_{u \in F} \Psi^{-1}\Psi\left(u(a_1^{(\delta_{u1})p_1})^*(a_2^{(\delta_{u2})p_2})^* \cdots (a_r^{(\delta_{ur})p_r})^*\right)$$

where

(a) $Y = \{a_1, a_2, \ldots, a_r\} \subseteq X$ *and* $\delta_u = (\delta_{u1}, \delta_{u2}, \ldots, \delta_{ur})$ *where* $\delta_{ui} \in \{0, 1\}$ *for any* $u \in F$ *and any* $i = 1, 2, \ldots, r$,

(b) $p_1, p_2, \ldots, p_r \geq 1$,

(c) F *is a finite language over* Y *with* $\epsilon \in F$ *satisfying*

(c)-(1) *there exist* $q_1, q_2, \ldots, q_r \geq 1$ *such that* $q_i \geq p_i$ *for any* $i = 1, 2, \ldots, r$ *and, for any* $u \in F$, *we have* $0 \leq c_i \leq q_i$, $i = 1, 2, \ldots, r$ *where* $\Psi(u) = (c_1, c_2, \ldots, c_r)$,

(c)-(2) *for any* $u, v \in F$, *there exists* $w \in F$ *such that*

$$\Psi(uv) \stackrel{\rightarrow}{\equiv} \Psi(w) \pmod{((\delta_{w1})p_1, (\delta_{w2})p_2, \ldots, (\delta_{wr})p_r)}$$

and $\delta_u, \delta_v \leq \delta_w$.

Proof $(\Rightarrow)$ Let $Y = alph(L) = \{a_1, a_2, \ldots, a_r\}$. Since L is regular, there exist $k_i, p_i \geq 1, i = 1, 2, \ldots, r$ such that $a_i^{k_i} \equiv a_i^{k_i+p_i}\ (P_L)$ for any $i = 1, 2, \ldots, r$. Let $q_i = k_i + p_i - 1, i = 1, 2, \ldots, r$ and let $F = \{u \in L \mid \Psi(u) = (c_1, c_2, \ldots, c_r),\ 0 \leq c_1 \leq q_1,\ 0 \leq c_2 \leq q_2, \ldots, 0 \leq c_r \leq q_r\}$.

Let $u \in F$ and let $i = 1, 2, \ldots, r$. Then let $\delta_{ui} = 0$ if $c_i < k_i$ and let $\delta_{ui} = 1$ if $c_i \geq k_i$ where $\Psi(u) = (c_1, c_2, \ldots, c_r)$. By $a_i^{k_i} \equiv a_i^{k_i+p_i}$ (P_L) for any $i = 1, 2, \ldots, r$, we have

$$\Psi^{-1}\Psi\left(u(a_1^{(\delta_{u1})p_1})^*(a_2^{(\delta_{u2})p_2})^* \cdots (a_r^{(\delta_{ur})p_r})^*\right) \subseteq L$$

for any $u \in F$. Now we assume that $v \in L \backslash F$. Let $\Psi(v) = (d_1, d_2, \ldots, d_r)$. Then there exists i, $1 \leq i \leq r$ such that $d_i > q_i$. Since $a_1^{d_1} a_2^{d_2} \cdots a_i^{d_i} \cdots a_r^{d_r} \in \Psi^{-1}\Psi(v) \subseteq L$ and $a_i^{k_i} \equiv a_i^{k_i+p_i}$ (P_L), we have $a_1^{d_1} a_2^{d_2} \cdots a_i^{d_i - mp_i} \cdots a_r^{d_r} \in L$ for some $m \geq 1$ where $k_i \leq d_i - mp_i \leq q_i$. Let $v_1 = a_1^{d_1} a_2^{d_2} \cdots a_i^{d_i - mp_i} \cdots a_r^{d_r}$. Then we have $v \in \Psi^{-1}\Psi(v_1(a_i^{p_i})^*) \subseteq L$. If $v_1 \in L \backslash F$, then we continue the same procedure. Finally, we have

$$v \in \Psi^{-1}\Psi\left(u(a_1^{(\delta_{u1})p_1})^*(a_2^{(\delta_{u2})p_2})^* \cdots (a_r^{(\delta_{ur})p_r})^*\right) \subseteq L$$

for some $u \in F$. This completes the proof that

$$L = \bigcup_{u \in F} \Psi^{-1}\Psi\left(u(a_1^{(\delta_{u1})p_1})^*(a_2^{(\delta_{u2})p_2})^* \cdots (a_r^{(\delta_{ur})p_r})^*\right).$$

Now let $u, v \in F$. If $uv \in F$, then obviously (c)-(2) holds. If $uv \notin F$, then by the above procedure

$$uv \in \Psi^{-1}\Psi\left(w(a_1^{(\delta_{w1})p_1})^*(a_2^{(\delta_{w2})p_2})^* \cdots (a_r^{(\delta_{wr})p_r})^*\right) \subseteq L$$

where $w \in F$. This means that $\Psi(uv) \rightrightarrows \Psi(w)$ $(\mathrm{mod}((\delta_{w1})p_1, (\delta_{w2})p_2, \ldots, (\delta_{wr})p_r)$. Moreover $\delta_u, \delta_v \leq \delta_w$ holds.

($\Leftarrow$) By Lemma 5.3.2, L is regular. It is obvious that L is commutative. Now we show that L is sh-closed. Let $\overline{u}, \overline{v} \in L$.

Then

$$\overline{u} \in \Psi^{-1}\Psi\left(u(a_1^{(\delta_{u1})p_1})^*(a_2^{(\delta_{u2})p_2})^* \cdots (a_r^{(\delta_{ur})p_r})^*\right)$$

and

$$\overline{v} \in \Psi^{-1}\Psi\left(v(a_1^{(\delta_{v1})p_1})^*(a_2^{(\delta_{v2})p_2})^* \cdots (a_r^{(\delta_{vr})p_r})^*\right)$$

where $u, v \in F$. Consequently, $\overline{u} \diamond \overline{v} \subseteq \Psi^{-1}\Psi(uv(a_1^{(\delta_{u1})p_1})^*(a_2^{(\delta_{u2})p_2})^*$

$\cdots(a_r^{(\delta_{ur})p_r})^*(a_1^{(\delta_{v1})p_1})^*(a_2^{(\delta_{v2})p_2})^*\cdots(a_r^{(\delta_{vr})p_r})^*)$. By (c)-(2), we have

$$\overline{u}\diamond\overline{v}\subseteq\Psi^{-1}\Psi\left(w(a_1^{(\delta_{w1})p_1})^*(a_2^{(\delta_{w2})p_2})^*\cdots(a_r^{(\delta_{wr})p_r})^*\right)$$

for some $w\in F$. Therefore, $\overline{u}\diamond\overline{v}\subseteq L$ and L is sh-closed.

Now consider a commutative sh-closed language whose sh-base is finite.

Proposition 5.3.5 *Let $L\subseteq X^*$ with $\epsilon\in L$. Then L is a commutative regular sh-closed language, whose sh-base is finite, if and only if L is represented as*

$$L=\bigcup_{u\in F}\Psi^{-1}\Psi\left(u(a_1^{p_1})^*(a_2^{p_2})^*\cdots(a_r^{p_r})^*\right)$$

where

(a) $Y=\{a_1,a_2,\ldots,a_r\}\subseteq X$,

(b) $p_1,p_2,\ldots,p_r\geq 1$,

(c) *F is a finite language over Y with $\epsilon\in F$ satisfying*

(c)-(1) *there exist $q_1,q_2,\ldots,q_r\geq 1$ such that $q_i\geq p_i$ for any $i=1,2,\ldots,r$ and for any $u\in F$ we have $0\leq c_i\leq q_i$, $i=1,2,\ldots,r$ where $\Psi(u)=(c_1,c_2,\ldots,c_r)$,*

(c)-(2) *for any $u,v\in F$ there exists $w\in F$ such that $\Psi(uv)\equiv\Psi(w)\ (\mathrm{mod}(p_1,p_2,\ldots,p_r))$.*

Proof ($\Rightarrow$) By Proposition 5.2.4, we can choose $k_i,p_i\geq 1$, $i=1,2,\ldots,r$ in the proof of Proposition 5.3.4 as those satisfying the conditions $a_i^{k_i}\equiv a_i^{k_i+p_i}\ (P_L)$ and $a_i^{p_i}\in L$ for any $i=1,2,\ldots,r$. In this case, since L is sh-closed, we have

$$L=\bigcup_{u\in F}\Psi^{-1}\Psi\left(u(a_1^{(\delta_{u1})p_1})^*(a_2^{(\delta_{u2})p_2})^*\cdots(a_r^{(\delta_{ur})p_r})^*\right)$$

$$=\bigcup_{u\in F}\Psi^{-1}\Psi(u(a_1^{p_1})^*(a_2^{p_2})^*\cdots(a_r^{p_r})^*).$$

($\Leftarrow$) The proof can be carried out in the same way as in Proposition 5.3.4 after assuming that $\delta_u = (1, 1, \ldots, 1, 1)$ for any $u \in F$, and after observing that the sh-base of L is contained in $F \cup \{a_1^{p_1}, a_2^{p_2}, \ldots, a_r^{p_r}\}$.

Proposition 5.3.6 *Let $L \subseteq X^*$ be a commutative sh-closed language whose sh-base $J(L)$ is finite. Then L is regular if and only if $a^+ \cap J(L) \neq \emptyset$ for any $a \in alph(L)$.*

Proof ($\Rightarrow$) This part has already been proven (see Proposition 5.2.4).

($\Leftarrow$) Let $\{a_1, a_2, \ldots, a_r\} = alph(L)$, let $\{a_1^{p_1}, a_2^{p_2}, \ldots, a_r^{p_r}\} \subseteq J(L)$ and let $J = \{u_1, u_2, \ldots, u_s\} = J(L) \setminus \{a_1^{p_1}, a_2^{p_2}, \ldots, a_r^{p_r}\}$ and let $\mu = \max\{|u| \mid u \in J\}$. Put $q = l.c.m.\{p_1, p_2, \ldots, p_r\}$ and $n = \mu q s$. Let $F = \bigcup_{x \in J^*,\, |x| \leq n} \Psi^{-1}\Psi(x)$. Obviously, F is a finite language. Now let $u \in L$. Then $u \in \Psi^{-1}\Psi\left(\alpha(a_1^{p_1})^*(a_2^{p_2})^* \cdots (a_r^{p_r})^*\right)$ where $\alpha \in J^*$, i.e. $\Psi(u) \overrightarrow{\equiv} \Psi(\alpha)(\mathrm{mod}(p_1, p_2, \ldots, p_r))$. If $|\alpha| > n$, then $\alpha \in \Psi^{-1}\Psi(\alpha' u_i^q)$ for some $\alpha' \in J^*$ and $i = 1, 2, \ldots, s$. It is obvious that $u_i^q \in \Psi^{-1}\Psi((a_1^{p_1})^*(a_2^{p_2})^* \cdots (a_r^{p_r})^*)$. Consequently, $u \in \Psi^{-1}\Psi(\alpha'(a_1^{p_1})^*(a_2^{p_2})^* \cdots (a_r^{p_r})^*)$, i.e. $\Psi(u) \overrightarrow{\equiv} \Psi(\alpha')((\mathrm{mod}(p_1, p_2, \ldots, p_r))$. We continue the same procedure if $|\alpha'| > n$ and we have $\beta \in J^*$ such that $|\beta| \leq n$ and $u \in \Psi^{-1}\Psi(\beta(a_1^{p_1})^*(a_2^{p_2})^* \cdots (a_r^{p_r})^*)$, i.e. $\Psi(u) \overrightarrow{\equiv} \Psi(\beta)(\mathrm{mod}(p_1, p_2, \ldots, p_r))$. Let $q_i = n$, $i = 1, 2, \ldots, r$. Then q_i, $i = 1, 2, \ldots, r$ satisfies the conditions (c)-(1) and (c)-(2) of Proposition 5.3.5. Hence L is regular.

5.4 Regular ssh-closed languages

In this section, we will determine the structure of regular ssh-closed languages.

Proposition 5.4.1 *Let $L \subseteq X^*$. Then L is a regular ssh-closed language if and only if L is represented as*

$$L = \bigcup_{u \in F} \Psi^{-1}\Psi\left(u(a_1^{p_1})^*(a_2^{p_2})^* \cdots (a_r^{p_r})^*\right)$$

where

(a) $Y = \{a_1, a_2, \ldots, a_r\} \subseteq X$,

(b) $p_1, p_2, \ldots, p_r \geq 1$,

(c) *F is a finite language with $\epsilon \in F$ satisfying*

(c)-(1) *for any $u \in F$ we have $0 \leq c_i \leq p_i$, $i = 1, 2, \ldots, r$ where $\Psi(u) = (c_1, c_2, \ldots, c_r)$,*

(c)-(2) *for any $u, v \in F$, there exists $w \in F$ such that $\Psi(uv) \overrightarrow{\equiv} \Psi(w)(\mathrm{mod}(p_1, p_2, \ldots, p_r)$, and*

(c)-(3) *for any $u, v \in F$, there exists $w \in F$ such that $\Psi(uw) \overrightarrow{\equiv} \Psi(v)(\mathrm{mod}(p_1, p_2, \ldots, p_r)$.*

Proof ($\Rightarrow$) First, notice that the ssh-base of L is a hypercode and hence finite (see the next section). From the fact that L is ssh-closed, it follows that $a_i^{p_i} \in L$ and $\epsilon \equiv a_i^{p_i}$ (P_L) for any $i = 1, 2, \ldots, r$ in the proof of Proposition 5.3.5. Therefore, we can assume that $k_i = 1$, i.e. $q_i = p_i$ for any $i = 1, 2, \ldots, r$. Therefore, it is enough to show (c)-(3).

Let $u, v \in F$ where $\Psi(u) = (c_1, c_2, \ldots, c_r)$ and let $\Psi(v) = (d_1, d_2, \ldots, d_r)$. If $d_i \geq c_i$ for any $i = 1, 2, \ldots, r$, then let $w \in \Psi^{-1}\{(d_1 - c_1, d_2 - c_2, \ldots, d_r - c_r)\}$. In this case, since $uw \in \Psi^{-1}\Psi(v) \subseteq L$ and L is ssh-closed, $w \in L$. On the other hand, $w \in F$ because $d_i - c_i \leq d_i \leq p_i$ for any $i = 1, 2, \ldots, r$. Now assume $c_i > d_i$ for some $i = 1, 2, \ldots, r$. Without loss of generality, we can assume that $c_1 > d_1, c_2 > d_2, \ldots, c_l > d_l$ and $d_{l+1} \geq c_{l+1}, d_{l+2} \geq c_{l+2}, \ldots, d_r \geq c_r$ for some $l = 1, 2, \ldots, r$. Notice that $d_1 + p_1 \geq c_1, d_2 + p_2 \geq c_2, \ldots, d_l + p_l \geq c_l, d_{l+1} \geq c_{l+1}, \ldots, d_r \geq c_r$. Let $w \in \Psi^{-1}(d_1+p_1-c_1, d_2+p_2-c_2, \ldots, d_l+p_l-c_l, d_{l+1}-c_{l+1}, \ldots, d_r-c_r)$. Then $uw \in \Psi^{-1}$ $(va_1^{p_1}a_2^{p_2} \cdots a_l^{p_l}) \subseteq L$. Since L is ssh-closed, $w \in L$. Moreover, $d_1 + p_1 - c_1 \leq p_1, d_2 + p_2 - c_2 \leq p_2, \ldots, d_l + p_l - c_l \leq p_l, d_{l+1} - c_{l+1} \leq p_{l+1}, \ldots, d_r - c_r \leq p_r$. Hence $w \in F$ and $\Psi(uw) \overrightarrow{\equiv} \Psi(v)$ $(\mathrm{mod}(p_1, p_2, \ldots, p_r))$.

($\Leftarrow$) It follows from Proposition 5.3.5 that L is a regular sh-closed language. Now let $u, v \in L$ and let $v = uw$ for some $w \in Y^*$. Since $u, v \in L$, there exist $u', v' \in F$ such that $u \in \Psi^{-1}\Psi(u'(a_1^{p_1})^*(a_2^{p_2})^* \cdots (a_r^{p_r})^*)$ and $v \in \Psi^{-1}\Psi$ $(v'(a_1^{p_1})^*(a_2^{p_2})^* \cdots (a_r^{p_r})^*)$. By (c)-(3), there exists $w' \in F$ such that $\Psi(u'w') \overrightarrow{\equiv} \Psi(v')$ $(\mathrm{mod}(p_1, p_2, \ldots, p_r))$, i.e. $u'w'$

$\in \Psi^{-1}\Psi(v'(a_1^{p_1})^*(a_2^{p_2})^* \cdots (a_r^{p_r})^*)$. Hence $uw' \in \Psi^{-1}\Psi(v'(a_1^{p_1})^*(a_2^{p_2})^* \cdots (a_r^{p_r})^*)$, i.e. $uw' \in \Psi^{-1}\Psi(v'(a_1^{p_1})^{k_1}(a_2^{p_2})^{k_2} \cdots (a_r^{p_r})^{k_r})$ for some $k_i \geq 0$, $i = 1, 2, \ldots, r$. On the other hand, $v \in \Psi^{-1}\Psi(v'(a_1^{p_1})^*(a_2^{p_2})^* \cdots (a_r^{p_r})^*)$ implies that $v \in \Psi^{-1}\Psi(v'(a_1^{p_1})^{t_1}(a_2^{p_2})^{t_2} \cdots (a_r^{p_r})^{t_r})$ for some $t_i \geq 0$, $i = 1, 2, \ldots, r$. Thus $uw'(a_1^{p_1})^{t_1} \cdots (a_r^{p_r})^{t_r} \in \Psi^{-1}\Psi(v(a_1^{p_1})^* \cdots (a_r^{p_r})^*)$. From $v = uw$, it follows that $w'(a_1^{p_1})^{t_1} \cdots (a_r^{p_r})^{t_r} \in \Psi^{-1}\Psi(w(a_1^{p_1})^*(a_2^{p_2})^* \cdots (a_r^{p_r})^*)$. Since $w' \in F$, $w \in \Psi^{-1}\Psi(w'(a_1^{p_1})^*(a_2^{p_2})^* \cdots (a_r^{p_r})^*)$. Thus $w \in L$. This means that L is ssh-closed.

5.5 Hypercodes

In this section, we will prove that every hypercode is finite. The proof is essentially due to [64] and [65].

On X^*, we have already defined an order relation, called the *embedding order* and denoted by $\leq_h$. For two words $u, v \in X^*$, we denote $u \leq_h v$ if there exists a $w \in X^*$ such that $v \in u \diamond w$. Recall that a language $H \subseteq X^+$ is called a *hypercode* if any distinct words $u, v \in H$ are incomparable with respect to the order $\leq_h$.

Let $u, v \in X^*$ and let $u \leq_h v$. Then v is called an *h-descendant* of u. By $hDes\{u\}$, we denote the set of all h-descendants of u.

Lemma 5.5.1 *Let $K \subseteq X^*$ be an infinite language. Assume that any hypercode contained in K is finite. Then there exists $w_i \in K, i \geq 1$ such that $w_{i+1} \in hDes(w_i)$ for any $i \geq 1$.*

Proof Let $K' = \{w'_1, w'_2, \ldots\}$ be an infinite subset of K such that $|w'_1| < |w'_2| < \cdots$ and let $H = \{w_i \in K' \mid K' \cap hDes(w_i) = \emptyset\}$. If $H = \emptyset$, then there exists an infinite subsequence $n_i, i \geq 1$ such that $n_1 = 1, w_{n_i} \in K'$ and $w_{n_{i+1}} \in K' \cap hDes(w_{n_i})$ for any $i \geq 1$. In this case, the lemma holds true. Assume $H \neq \emptyset$. Then $H \subseteq X^*$ is a hypercode. Thus $|H| < \infty$ and there exists a positive integer k_0 such that $K' \cap hDes(w_i) \neq \emptyset$ for any $i \geq k_0$. Hence there exists an infinite subsequence $n_i, i \geq 1$ such that $n_1 = k_0, w_{n_i} \in K'$ and $w_{n_{i+1}} \in K' \cap hDes(w_{n_i})$ for any $i \geq 1$. Therefore, the lemma holds true. This completes the proof of the lemma.

Now we are ready to prove that any hypercode is finite.

Proposition 5.5.1 *Let $H \subseteq X^*$ be a hypercode. Then H is finite.*

Proof Let $|X| = 1$, i.e. $X = \{a_1\}$. Then it is obvious that any hypercode over X is finite. Now assume that the proposition holds true for X with $|X| < n$. Suppose that the proposition does not hold true for X with $|X| = n$. Let $X = \{a_1, a_2, \ldots, a_{n-1}, a_n\}$. Then there exists an infinite hypercode $H \subseteq X^*$. Let $\mathcal{H} = \{H \subseteq X^* \mid H$ is a hypercode such that $|H| = \infty\}$. Let $u \in H$ for some $H \in \mathcal{H}$ such that $|u| = min\{|v| \mid \exists H \in \mathcal{H}, v \in H\}$.

If $u = b \in X$, then $H \setminus \{u\}$ is a hypercode over $X \setminus \{b\}$. Thus by the induction hyposis, $H \setminus \{u\}$ is finite and hence H is finite, which is a contradiction.

Assume that $u = b_1 b_2 \cdots b_r$ where $r \geq 2$ and $b_i \in X$ for any $i = 1, 2, \ldots, r$. Let $v = b_1 b_2 \cdots b_{r-1}$. Consider the set $D = \{w \mid w \in H, v \leq_h w\}$. If D is a finite set, then $(H \setminus D) \cup \{v\}$ is an infinite hypercode, which contradicts the minimality of $|u|$. Hence D is an infinite hypercode. Let $\{w_1, w_2, \ldots\}$ be an infinite subset of D such that $|w_1| < |w_2| < \cdots$. Then for any $i \geq 1$, w_i can be represented as $w_i = u_{i1} b_1 u_{i2} b_2 \cdots b_{i(r-2)} u_{i(r-1)} b_{r-1} u_{ir}$ where $u_{i1} \in (X \setminus \{b_1\})^*, u_{i2} \in (X \setminus \{b_2\})^*, \ldots, u_{i(r-1)} \in (X \setminus \{b_{r-1})^*\}$ and $u_{ir} \in (X \setminus \{b_r\})^*$. By the induction hypothesis, for each $i \geq 1$, a hypercode contained in $\{u_{it} \mid t \geq 1\}$ is finite. By Lemma 5.5.1, there exists an infinite subset $\{w'_i \mid i \geq 1\}$ of $\{w_1, w_2, \ldots\}$ such that $w'_i = u'_{i1} b_1 u'_{i2} b_2 \cdots b_{r-2} u'_{i(r-1)} b_{r-1} u'_{ir}, i \geq 1$ and, $u'_{it} \leq_h u_{(i+1)t}$ for any $i \geq 1$ and any $t = 1, 2, \ldots, r$. However, this contradicts the assumption that H is a hypercode. Consequently, $|H| < \infty$ if $H \subseteq X^*$ is a hypercode. This completes the proof of the proposition.

Chapter 6

Insertions and Deletions

In Chapter 4, we dealt with n-insertions, shuffle operations and related languages. In this chapter, we will consider the *insertion* operation, i.e. 1-insertion operation, and its inverse operation, i.e. *deletion* operation in details. First we will introduce the notions of *insertion residual* and *insertion closure* of a language and we will provide several properties with respect to these notions. Then we will introduce the notion of *deletion closure* of a language. In Chapter 5, we determined the structure of commutative regular shuffle closed languages. In this chapter, we will provide some properties of regular insertion closed languages. Finally, we will also deal with a problem related to the combination of two operations, i.e. insertion closure and insertion residual operations.

The results in this section are due to [40].

6.1 Insertion closures

First we recall that the insertion of u into v is the language $u \rhd v = \{v_1uv_2 \mid v = v_1v_2, v_1, v_2 \in X^*\}$. Notice that, in Chapter 4, we denoted $u \rhd^{[1]} v$ instead of $u \rhd v$. Let $L \subseteq X^*$. Then the *insertion residual* $inRes(L)$ of L is defined as : $inRes(L) = \{x \in X^* \mid \forall u \in L, x \rhd u \subseteq L\}$.

Example 6.1.1 Let $X = \{a, b\}$. Then (1) $inRes(X^*) = X^*$. (2) $inRes(L_{ab}) = L_{ab}$ where $L_{ab} = \{x \in X^* \mid |x|_a = |x|_b\}$. (3) $inRes(L)$

$= \{\epsilon\}$ for $L = \{a^n b^n \mid n \geq 0\}$. (4) $inRes(L_1) = L_1$ and $inRes(L_2) = L_1$ for $L_1 = (a^2)^*$ and $L_2 = aL_1$. (5) $inRes(L) = b^*$ for $L = b^*ab^*$. (6) $inRes(L) = L$ for $L = aX^*b$.

For a commutative language, we have the following result.

Proposition 6.1.1 *Let $L \subseteq X^*$. Then $inRes(L)$ is a submonoid of X^*. Moreover, if L is a commutative language, then $inRes(L)$ is also a commutative language.*

Proof Let $x, y \in inRes(L)$ and let $u = u_1u_2 \in L$. Then $u_1xu_2 \in L$ and hence $u_1xyu_2 \in L$. Therefore, $xy \in inRes(L)$. Since $\epsilon \in inRes(L)$, $inRes(L)$ is a monoid. To prove that $inRes(L)$ is commutative, it is enough to show that $xuvy \in inRes(L)$ implies $xvuy \in inRes(L)$. If $w \in L$ and $w = w_1w_2$, then $w_1xuvyw_2 \in L$ and hence $w_1xvuyw_2 \in L$. Therefore, $xvuy \in inRes(L)$.

Now recall the notion of insertion. Let L_1, L_2 be two languages over X. The *insertion* of L_2 into L_1 is defined as: $L_1 \triangleright L_2 = \{u_1vu_2 \mid u_1u_2 \in L_2, v \in L_1\}$.

The iterated insertion can also be defined as follows:

$$L_1 \triangleright^* L_2 = \bigcup_{n=0}^{\infty} (L_1 \triangleright^n L_2)$$

where $L_1 \triangleright^0 L_2 = L_1$ and $L_1 \triangleright^{i+1} L_2 = L_1 \triangleright (L_1 \triangleright^i L_2)$ For any $i \geq 0$.

Lemma 6.1.1 *Let $L \subseteq X^*$ and let $u, v \in inRes(L)$. Then $v \triangleright^* u \subseteq inRes(L)$.*

Proof Let $w \in (v \triangleright^* u)$. There exists $k \geq 0$ such that $w \in (v \triangleright^k u)$. We will show, by the induction on k, that $w \in inRes(L)$. If $k = 0$, then $w = v \in inRes(L)$. Assume the assertion holds true for k and take $w \in (v \triangleright^{k+1} u)$ and $z = z_1z_2 \in L$. Then $w = w_1vw_2$ where $w_1w_2 \in (v \triangleright^k u) \subseteq inRes(L)$. Consequently, $z_1w_1w_2z_2 \in L$. This, together with the fact that $v \in inRes(L)$ imply that $z_1w_1vw_2z_2 \in L$. Since $z = z_1z_2$ was an arbitrary word in L, we deduce that $w \in inRes(L)$.

Proposition 6.1.2 *Let $L \subseteq X^*$. Then we have $inRes^2(L) = inRes$ $(inRes(L)) = inRes(L)$.*

Proof Assume $u \in inRes(inRes(L))$. As $\epsilon \in inRes(L)$, we have $u = \epsilon u \in inRes(L)$, i.e. $inRes(inRes(L)) \subseteq inRes(L)$. Assume now that $v \in inRes(L)$. Let $u = u_1u_2 \in inRes(L)$. Consider $u_1vu_2\, inX^*$. Obviously, $v_1uv_2 \in v \rhd^* u$. By Lemma 6.1.1, $u_1vu_2 \in inRes(L)$ and hence $v \in inRes(inRes(L))$, i.e. $inRes(L) \subseteq inRes(inRes(L))$.

Proposition 6.1.3 *If a language L is regular, then $inRes(L)$ is regular.*

Proof Let $u, v \in X^*$ and let $u \equiv v(P_L)$. Assume $xuy \in inRes(L)$ for $x, y \in X^*$. Let $w = w_1w_2 \in L$. Then $w_1xuyw_2 \in L$ and hence $w_1xvyw_2 \in L$. Thus $xvy \in inRes(L)$. Symmetrically, $xvy \in inRes(L)$ implies $xuy \in inRes(L)$. This means that $u \equiv v(P_{inRes(L)})$. Consequently, the number of congruence classes of $P_{inRes(L)}$ is finite and hence $inRes(L)$ is regular.

A language L such that $L \subseteq inRes(L)$ is called *insertion closed* (or in short, *ins-closed*). A language L is ins-closed if and only if $u = u_1u_2 \in L$ and $v \in L$ imply $u_1vu_2 \in L$. Consequently, every ins-closed language is a subsemigroup of X^*.

Notice that a language L is ins-closed if and only if $L \rhd L \subseteq L$, and hence $L \rhd^* L \subseteq L$. Therefore, $L \rhd^* L$ is an ins-closed language for any language $L \subseteq X^*$ and $L \rhd^* L$ is the smallest ins-closed language containing L, which is called the *inserttion closure* of L.

Theorem 6.1.1 *Let $|X| \geq 2$. Then we have: (i) If L is context-free, then the insertion closure of L is context-free. (ii) If L is regular, then the insertion closure of L is not always regular.*

Proof (i) Let $L \subseteq X^*$ be context-free and let $\mathcal{G} = (V, X, P, S)$ be a context-free grammar with $L = \mathcal{L}(\mathcal{G})$. Moreover, let $\rho(A) = A$ for any $A \in V$ and $\rho(a) = [a]$ for any $a \in X \cup \{\epsilon\}$. Now we construct the context-free grammar $\overline{G} = (\overline{V}, X, \overline{P}, S)$ as follows: (1) $\overline{V} = V \cup \{[a] \mid a \in X \cup \{\epsilon\}\}$. (2) $\overline{P} = \{A \to \rho(\alpha_1)\rho(\alpha_2)\cdots\rho(\alpha_k) \mid (A \to \alpha_1\alpha_2\cdots\alpha_k) \in P\} \cup \{[a] \to [a]S, [a] \to S[a] \mid a \in X \cup \{\epsilon\}\} \cup \{[a] \to a$

$\mid a \in X \cup \{\epsilon\}\}$. Then it can be easily verified that the grammar $\mathcal{G}$ generates the insertion closure of L. Consequently, the insertion closure of L is context-free. (ii) Let $X = \{a, b, \ldots\}$ and let $L = \{ab\}$. Then L is regular. However, since $(L \rhd^* L) \cap a^+b^+ = \{a^n b^n \mid n \geq 1\}$, the insertion closure of L is not regular.

Remark 6.1.1 Let L be a context-free language which is accepted by a pushdown acceptor $\boldsymbol{A}$. Making use of $\boldsymbol{A}$, we can construct the pushdown acceptor which accepts the insertion closure of L. We leave this problem to the readers.

6.2 Deletion closures

We start this section with providing some operations on words and languages.

Definition 6.2.1 Let $u, v \in X^*$. Then $u \longrightarrow v = \{u_1u_2 \mid u = u_1vu_2\}$ is called the *deletion* of v from u. For languages $L_1, L_2 \subseteq X^*$, $L_1 \longrightarrow L_2 = \{u_1u_2 \in X^* \mid u_1vu_2 \in L_1, v \in L_2\}$ is called the *deletion* of L_2 from L_1. Moreover, we can define the *iterated deletion* of L_2 from L_1 as follows: $L_1 \longrightarrow^* L_2 = \bigcap_{n=0}^{\infty}(L_1 \longrightarrow^n L_2)$ where $L_1 \longrightarrow^0 L_2 = L_1$ and $L_1 \longrightarrow^{i+1} L_2 = (L_1 \longrightarrow^i L_2) \longrightarrow L_2$ for any $i \geq 0$. For $u, v \in X^*$, $u \longrightarrow^* v$ means $\{u\} \longrightarrow^* \{v\}$.

Proposition 6.2.1 *The deletion of a regular language from a regular language is regular.*

Proof Let $L, K \subseteq X^*$ be regular languages. Let $\boldsymbol{A} = (S, X, \delta, s_0, F)$ and $\boldsymbol{B} = (T, X, \gamma, t_0, G)$ be finite acceptors such that $\mathcal{T}(\boldsymbol{A}) = L$ and $\mathcal{T}(\boldsymbol{B}) = K$, respectively. Moreover, let $X' = \{a' \mid a \in X\}$.

Consider the following finite acceptors $\boldsymbol{A}' = (S, X \cup X', \delta', s_0, F)$ and $\boldsymbol{B}' = (T, X \cup X', \gamma', t_0, G)$: (1) $sa^{\boldsymbol{A}'} = sa'^{\boldsymbol{A}'} = sa^{\boldsymbol{A}}$ for any $s \in S$ and any $a \in X$. (2) $ta^{\boldsymbol{B}'} = t$ for any $t \in T$ and any $a \in X$, and $ta'^{\boldsymbol{B}'} = ta^{\boldsymbol{B}}$ for any $t \in T$ and any $a' \in X'$.

Let ρ be the following homomorphism of $(X \cup X')^*$ onto X^*: $\rho(a) = a$ and $\rho(a') = \epsilon$ for any $a \in X$.

Now we prove that $\rho(\mathcal{T}(\boldsymbol{A}') \cap \mathcal{T}(\boldsymbol{B}') \cap X^*X'^*X^*) = L \longrightarrow K$.

Let $u \in \rho(\mathcal{T}(\boldsymbol{A}') \cap \mathcal{T}(\boldsymbol{B}') \cap X^*X'^*X^*)$. Then there exist $v' \in X'^*$ and $u_1, u_2 \in X^*$ such that $u = \rho(u_1v'u_2) = u_1u_2$ and $u_1v'u_2 \in \mathcal{T}(\boldsymbol{A}') \cap \mathcal{T}(\boldsymbol{B}')$. Let θ be the homomorphism of $(X \cup X')^*$ onto X^* such that $\theta(a) = \theta(a') = a$ for any $a \in X$. Thus $u_1\theta(v')u_2 \in \mathcal{T}(\boldsymbol{A}) = L$ and $\theta(v') \in \mathcal{T}(\boldsymbol{B})$. Hence $u = u_1u_2 \in L \longrightarrow K$.

On the other hand, let $u = u_1u_2 \in L \longrightarrow K$. Then there exists $v \in X^*$ such that $u_1vu_2 \in L$ and $v \in K$. Hence there exists $v' \in X'^*$ such that $v = \theta(v')$. Consequently, $u_1v'u_2 \in \mathcal{T}(\boldsymbol{A}') \cap \mathcal{T}(\boldsymbol{B}') \cap X^*X'^*X^*$. This means that $u = u_1u_2 = \rho(u_1v'u_2) \in \rho(\mathcal{T}(\boldsymbol{A}') \cap \mathcal{T}(\boldsymbol{B}') \cap X^*X'^*X^*)$, i.e. $\rho(\mathcal{T}(\boldsymbol{A}') \cap \mathcal{T}(\boldsymbol{B}') \cap X^*X'^*X^*) = L \longrightarrow K$.

Since $\mathcal{T}(\boldsymbol{A}') \cap \mathcal{T}(\boldsymbol{B}') \cap X^*X'^*X^*$ is regular, $L \longrightarrow K$ is regular.

It is noticed that the above proposition does not hold for context-free languages. For instance, let $X = \{a, b, c, d\}$, let $L = \{a^nb^nc^md^m \mid n, m \geq 1\}$ and let $K = \{b^kc^k \mid k \geq 1\}$. By the iteration property for context-free languages, $L \longrightarrow K = \{a^nb^{n-k}c^{m-k}d^m \mid n, m \geq k \geq 1\}$ is not context-free though L and K are context-free.

Let $L \subseteq X^*$ and let $Sub(L) = \{u \in X^* \mid xuy \in L\}$, i.e. $Sub(L)$ is the set of the subwords of the words in L. Then, by $del(L)$ we denote the following language: $del(L) = \{x \in Sub(L) \mid$ for any $u \in L, u = u_1xu_2$ implies $u_1u_2 \in L\}$.

Example 6.2.1 Let $X = \{a, b\}$. Then (1) $del(X^*) = X^*$. (2) $del(L_{ab}) = L_{ab}$ where $L_{ab} = \{u \in X^* \mid |u|_a = |u|_b\}$. (3) $del(L) = L$ for $L = \{a^nb^n \mid n \geq 0\}$. (4) $del(L) = b^*$ for $L = b^*ab^*$.

Proposition 6.2.2 *Let $L \subseteq X^*$. Then we have: (i) If $x, y \in del(L)$ and $xy \in Sub(L)$, then $xy \in del(L)$. (ii) If $Sub(L)$ is a submonoid of X^*, then $del(L)$ is a submonoid of X^*. (iii) If L is a commutative language, then $del(L)$ is also commutative.*

Proof (i) Let $x, y \in del(L)$ with $xy \in Sub(L)$. If $u = u_1xyu_2 \in L$, then $u_1yu_2 \in L$ and consequently $u_1u_2 \in L$. Therefore $xy \in del(L)$.

(ii) Let $x, y \in del(L)$. Then $x, y \in Sub(L)$ and $xy \in Sub(L)$. By (i), $xy \in del(L)$. Since $\epsilon \in del(L)$, $del(L)$ is a submonoid of X^*.

(iii) It is enough to show that $xuvy \in del(L)$ implies $xvuy \in del(L)$. Since L is commutative, $u_1xuvyu_2 \in L$ if and only if $u_1xvuyu_2 \in L$. As $xuvy \in del(L)$, we have $u_1u_2 \in L$. This implies that $xvuy \in del(L)$ as well.

A language L is called *deletion closed* (or in short, *del-closed*) if $v \in L$ and $u_1vu_2 \in L$ imply $u_1u_2 \in L$. For example, X^* and L_{ab} are del-closed languages that are also ins-closed. Notice that both languages are submonoids of X^*.

The notion of a del-closed language is strongly connected to the operations of deletion and iterated deletion.

Notice that a language $L \subseteq X^*$ is del-closed if and only if $L \longrightarrow L \subseteq L$.

Proposition 6.2.3 *Let $L \subseteq X^*$ be an ins-closed language. Then L is del-closed if and only if $L = L \longrightarrow L$.*

Proof ($\Rightarrow$) Since L is del-closed, $L \longrightarrow L \subseteq L$. Now let $u \in L$. Since L is ins-closed, $uu \in L$. Therefore $u \in L \longrightarrow L$, i.e. $L \subseteq L \longrightarrow L$. Hence $L = L \longrightarrow L$.

($\Leftarrow$) Since $L \longrightarrow L \subseteq L$, L is del-closed.

If L is a nonempty language and if D_L is the family of all del-closed languages L_i containing L, then the language

$$\bigcap_{L_i \in D_L} L_i$$

is a del-closed language called the *deletion closure* (or in short, *del-closure*) of L. The del-closure of L is the smallest del-closed language containing L.

We will now define a sequences of languages whose union is the del-closure of a given language L.

$$D_0(L) = L$$

$$D_1(L) = D_0(L) \longrightarrow (D_0(L) \cup \{\epsilon\})$$

$$D_2(L) = D_1(L) \longrightarrow (D_1(L) \cup \{\epsilon\})$$

$$\cdots$$

$$D_{k+1}(L) = D_k(L) \longrightarrow (D_k(L) \cup \{\epsilon\})$$

$$\cdots$$

Clearly $D_k(L) \subseteq D_{k+1}(L)$. Let

$$D(L) = \bigcup_{k \geq 0} D_k(L).$$

Proposition 6.2.4 *$D(L)$ is the del-closure of the language L.*

Proof Clearly $L \subseteq D(L)$. Let $v \in D(L)$ and $u_1vu_2 \in D(L)$. Then $v \in D_i(L)$ and $u_1vu_2 \in D_j(L)$ for some integers $i, j \geq 0$. If $k = max\{i, j\}$, then $v \in D_k(L)$ and $u_1vu_2 \in D_k(L)$. This implies $u_1u_2 \in D_{k+1}(L) \subseteq D(L)$. Therefore $D(L)$ is a del-closed language containing L.

Let M be a del-closed language containing $L = D_0(L)$. Since M is del-closed, if $D_k(L) \subseteq M$ then $D_{k+1}(L) \subseteq M$. Hence $D_i(L) \subseteq M$ for any $i \geq 1$. Thus $D(L) \subseteq M$ and $D(L)$ is the del-closure of L.

For the class of regular languages, we have the following result.

Theorem 6.2.1 *If $L \subseteq X^*$ is regular, then $D(L)$ is regular.*

Proof First, consider the case $|X| = 1$, i.e. $X = \{a\}$. If L is finite, then $D(L)$ is finite and hence regular. Assume L is infinite. Let $a^n \in L$ for some $n \geq 1$. Let $M = \{a^i \mid 0 \leq i \leq n, |a^i(a^n)^* \cap L| = \infty\}$. Then it can be seen that $D(L) = M(a^n)^* \cup F$ where $F \subseteq X^*$ is finite. Consequently, $D(L)$ is regular.

Now consider the case $|X| \geq 2$. We can assume that a regular language $L \subseteq X^*$ contains the empty word without loss of generality. Let $\boldsymbol{A} = (S, X, \delta, S_0, F)$ be a nondeterministic acceptor accepting L, i.e. $D_0(L)$. For any $(t, a) \in S \times X$, let $\delta_1(t, a) = \delta(t, a) \cup \{q \in S \mid \exists p \in \delta(t, a), \exists u \in L, q \in \delta(p, u)\}$. Moreover, let $S_1 = S_0 \cup \{q \in S \mid \exists p \in S_0, \exists u \in L, q \in \delta(p, u)\}$. Now let $\boldsymbol{A}_1 = (S, X, \delta_1, S_1, F)$. Then $\boldsymbol{A}_1$ can be effectively constructed and $D_1(L) \subseteq \mathcal{T}(\boldsymbol{A}_1) \subseteq D(L)$ where $\mathcal{T}(\boldsymbol{A}_1)$ is the language accepted by $\boldsymbol{A}_1$. In the same way, we can construct a nondeterministic acceptor $\boldsymbol{A}_2$ with the same state set S

such that $D_2(L) \subseteq \mathcal{T}(\boldsymbol{A}_2) \subseteq D(L)$. Continuing the same procedure, for any $n \geq 3$ we can construct a nondeterministic acceptor $\boldsymbol{A}_n$ with the state set S such that $D_n(L) \subseteq \mathcal{T}(\boldsymbol{A}_n) \subseteq D(L)$. By the finiteness of S, this procedure stops after a finite number of steps m and we have $D_m(L) = \mathcal{T}(\boldsymbol{A}_m) = D(L)$. This means that $D(L)$ is regular.

However, for the class of context-free languages, the situation is different. Before describing the situation, we provide the notion of codes though at the moment we need only the notion of infix codes.

Let $C \subseteq X^+$ be a language over X. Moreover, let $u_1, u_2, \ldots, u_r, v_1, v_2, \ldots, v_s \in C$ where r and s are positive integers. If $u_1 u_2 \cdots u_r = v_1 v_2 \cdots v_s$ implies that $r = s$ and $u_i = v_i$ for any $i = 1, 2, \ldots, r$, then C is called a *code* over X. A typical code is a prefix (suffix) code. A code $C \subseteq X^+$ is called a *prefix code* (*suffix code*) if $u, uv \in C$ ($u, vu \in C$) implies that $v = \epsilon$, i.e. $C \cap CX^+ \neq \emptyset$ ($C \cap X^+C \neq \emptyset$). A code $C \subseteq X^+$ is called a *bifix code* if C is a prefix code and at the same time a suffix code. A code $C \subseteq X^+$ is called a *maximal code* if $C \cup \{u\}$ is not a code for any $u \in X^+ \setminus C$. In the same way, a prefix (suffix) code $C \subseteq X^+$ is called a *maximal prefix code* (*maximal suffix code*) if $C \cup \{u\}$ is not a prefix (suffix) code for any $u \in X^+ \setminus C$. It is well known that a maximal prefix (suffix) code is a maximal code. Moreover, a prefix (suffix) code $C \subseteq X^+$ is maximal if and only if $u \in CX^*$ ($u \in X^*C$) or $C \cap uX^* \neq \emptyset$ ($C \cap X^*u \neq \emptyset$) for any $u \in X^*$.

Let $C \subseteq X^+$. Then C is called an *infix code* if $v, uvw \in C$ implies that $uw = \epsilon$. It is obvious that an infix code is a bifix code and hence a code.

Proposition 6.2.5 *If $|X| \geq 2$, then there exists a context-free language $L \subseteq X^*$ such that $D(L)$ is not context-free.*

Proof First, assume $|X| \geq 4$. Let $X = \{\$, a, b, c, \ldots\}$. Consider the following language:

$$L = \{\$a^{i_1}cb^{2i_1}a^{i_2}cb^{2i_2}\cdots a^{i_k}cb^{2i_k}\$ \mid k \geq 1, i_t \geq 1, 1 \leq t \leq k\} \cup \{ba\}.$$

Since the set $\{\$u\$ \mid u \in (X \setminus \{\$\})^*\}$ is an infix code [65], $D(L) = L \rightarrow^* \{ba\}$ and hence $D(L) \cap \$ac^+b^+\$ = \{\$ac^kb^{2^k}\$ \mid k \geq 1\}$. Therefore, $D(L)$ is not context-free though L is context-free.

For the general case, i.e. $X = \{a, b, \ldots\}$, consider the following coding for the letters $\$, a, b$ and c in the above:

$$\$ \to bbbbb,\ a \to abaaa,\ b \to abbaa,\ c \to abbba.$$

Moreover, consider the following language:

$M = \{(bbbbb)(abaaa)^{i_1}(abbba)(abbaa)^{2i_1}(abaaa)^{i_2}(abbba)$ $(abbaa)^{2i_2} \cdots (abaaa)^{i_k}(abbba)(abbaa)^{2i_k}(bbbbb) \mid k \geq 1, i_t \geq 1,$ $t = 1, 2, \ldots, k\} \cup \{(abbaa)(abaaa)\}$.

Then $\{(bbbbb)u(bbbbb) \mid u \in \{abaaa, abbaa, abbba\}^*\}$ becomes an infix code over X. On the other hand, since $\{(abbaa)(abaaa)\}$ does not contain any word in the language $\{(bbbbb)u(bbbbb) \mid u \in \{abaaa, abbaa, abbba\}^+\}$ as a subword, $D(M) = (M \longrightarrow^* \{(abbaa)(abaaa)\})$. Moreover, since $(abbaa)(abaaa)$ can be found only at the position between the suffix of $(abbaa)^i$ and the prefix of $(abaaa)^j$, $D(M) \cap (bbbbb)(abaaa)(abbba)^+(abbaa)^+(bbbbb) = \{(bbbbb)(abaaa)(abbba)^k(abbaa)^{2^k}(bbbbb) \mid k \geq 1\}$. Therefore, $D(M)$ is not context-free though M is context-free.

6.3 Ins-closed and del-closed languages

Let $L \subseteq X^*$ be a nontrivial ins-closed language and let $J = (L \setminus \{\epsilon\}) \setminus ((L \setminus \{\epsilon\}) \rhd (L \setminus \{\epsilon\}))$. Then $L \setminus \{\epsilon\} = J \rhd^* J$ and J is called the *insertion base* (or in short, *ins-base*) of L.

The following result shows that if L is regular, its ins-base is also regular.

Proposition 6.3.1 *If L is a regular ins-closed language, then its ins-base J is a regular language.*

Proof Since L is regular, $J = (L \setminus \{\epsilon\}) \setminus ((L \setminus \{\epsilon\}) \rhd (L \setminus \{\epsilon\}))$ is regular.

The ins-base of a regular ins-closed language can be an infinite language. For example, $L = ba^*b \rhd^* ba^*b = \{bxb \mid x \in X^*, |x|_b \text{ is even}\}$ is regular and ins-closed but J contains the infinite set ba^*b. If we put the additional constraint that L is del-closed, the ins-base will always be finite, as will be shown by Proposition 6.3.3.

Proposition 6.3.2 *Let $L \subseteq X^*$ be a regular language. If L is ins-closed and del-closed, then the minimal set of generators K of L, i.e. $K = (L \backslash \{\epsilon\}) \backslash (L \backslash \{\epsilon\})^2$ is a regular maximal code over alph(L). In fact, K is a maximal prefix code and a maximal suffix code over alph(L).*

Proof Since $K = (L \backslash \{\epsilon\}) \backslash (L \backslash \{\epsilon\})^2$, K is regular. Moreover, since L is del-closed, K is a bifix code. Now we prove that K is a maximal prefix (suffix) code over *alph*(L).

Let $a \in alph(L)$. Then $uav \in L$ for some $u, v \in alph(L)^*$. Therefore, $u^n(av)^n \in L$ for any $n \geq 1$. We take an enough big integer as n, e.g. $n > |S|$ where S is the set of states of a finite acceptor accepting L. Then there exists $i = 1, 2, \ldots, |S|$ such that $u^{n+i}(av)^n \in L$. On the other hand, $u^n(av)^n \in L$. Since L is del-closed, $u^i \in L$. As $u^i \in L$, we have $(av)^i \in L$, i.e. $aw \in L$ where $w = (va)^{i-1}v$. In the same way as above, there exists a positive integer i_a such that $a^{i_a} \in L$.

Let now $x \in X^+$. Then $x = a_1 a_2 \cdots a_r$ where $a_i \in X, i = 1, 2, \ldots, r$. Let $H = \{a_1^{i_1}, a_2^{i_2}, \ldots, a_r^{i_r}\} \subseteq L$. Obviously, $xy = a_1 a_2 \cdots a_r y \in (H \rhd^* H) \subseteq L$ for some $y \in X^*$. This means that K is a maximal prefix code. The proof that K is a maximal suffix code over *alph*(L) can be done in the same way. Hence K is a maximal bifix code over *alph*(L) as well.

In the above proof, the condition of regularity is necessary.

Remark 6.3.1 Let $X = \{a, b\}$ and let $L = (ab \rhd^* ab) \cup \{\epsilon\}$. Then L is ins-closed and del-closed. Moreover, K is a bifix code. But K is not a maximal bifix code over $\{a, b\}$ since $\{ba\} \cup K$ is a bifix code where K is the minimal set of generators of L.

The converse of the above proposition does not hold.

Remark 6.3.2 Let $X = \{a, b\}$ and let $L = \{a^2\} \cup \{b^2\} \cup ab^+a \cup ba^+b$. Then L is a regular maximal bifix code over $\{a, b\}$. But L^* is not ins-closed since $ab(aba)a = ababaa \notin L^*$ though $aba \in L$.

Lemma 6.3.1 *Let $L \subseteq X^*$ be a regular language that is ins-closed and del-closed. Then there exists a positive integer n such that for every $u \in X^+$ we have $u^n \in L$.*

Proof Since the minimal set of generators of L is a maximal prefix code, for any $u \in alph(L)^+$ there exists $y \in X^*$ such that $uy \in L$. Let $p \geq 1$ be an integer. Then $u^p y^p \in L$. In the same way as in the proof of Proposition 6.3.2, $u^i \in L$ for some positive integer $i \leq |S|$ where S is the set of states of a finite acceptor accepting L. Let $n = |S|!$. Then $u^n \in L$.

Proposition 6.3.3 *Let $L \subseteq X^*$ be a regular language that is ins-closed and del-closed. Let J be the ins-base of L. Then J is finite.*

Proof Suppose J is infinite. Then by the iteration property for a regular language, there exist $u, v, w \in alph(L)^+$ such that $uvw \in J$ and $uv^*w \subseteq L$. Hence $uw \in L$. By Lemma 6.3.1, there exists n, $n \geq 1$ such that $z^n \in L$ for any $z \in alph(L)^*$. Since L is ins-closed, $uv(u^{n-1}w^{n-1})w = uvu^{n-1}w^n \in L$. From the assumption that L is del-closed, it follows that $uvu^{n-1} \in L$. On the other hand, since $u^n \in L$, $u^{n-1}(uvu^{n-1})u = u^n v u^n \in L$. By the assumption that L is del-closed, $v \in L$. However, this contradicts the assumption that $uvw \in J$. Therefore, J must be finite.

6.4 Combination of operations

In this section, we will combine the operations on languages. The main result is due to [44].

First we consider $inRes(u \rhd^* u)$ for $u \in X^+$.

Lemma 6.4.1 *Let $u, v \in X^+$ and let $v \in inRes(u \rhd^* u)$. If $|u| = |v|$, then there exists $p \in X^+$ such that $u, v \in pX^*$.*

Proof Obviously, the lemma holds if $u = v$. Now let $u \neq v$. Since $vu \in inRes(u \rhd^* u)$ and $u \neq v$, there exist $p, q \in X^+$ such that $u = pq$, $vu = puq$. Hence $u, v \in pX^*$.

For $u \in X^*$, we define $Int(u)$ as follows: $Int(\epsilon) = \epsilon$ and $Int(au') = a$ if $a \in X$ and $u' \in X^*$.

Let $u \in X^+$ and let $i \geq 1$. Then we define $u(i) = a$ if $u = u'au'', u', u'' \in X^*, a \in X$ and $|u'a| = i$.

Remark 6.4.1 Let $u, v \in X^+$ and let $v \in inRes(u \triangleright^* u)$. If $|u| = |v|$ and $u \neq v$, then there exists $p \in X^+$ such that $u = pu', v = pv'$ and $Int(u') \neq Int(v')$.

Proposition 6.4.1 *Let $u, v \in X^+$ and $v \in inRes(u \triangleright^* u)$. If $|u| = |v|$, then $u = v$.*

Proof Suppose $u \neq v$. By Remark 6.4.1, $u = pu', v = pv'$ and $Int(u') \neq Int(v')$ where $p \in X^+$ and $u', v' \in X^*$.

Case A $p = a^i bp', i \geq 1, a, b \in X, a \neq b$ and $p' \in X^*$.

Consider $w = a^i a^i bp'v'bp'u' \in u \triangleright^* u$. Then $w = \alpha u \beta$ where $\alpha\beta = u$ and $|\alpha| \leq |u|$. Since $u \in a^i bX^*$, $|\alpha| \leq i$. If $|\alpha| = i$, then $u = v$, which is a contradiction. Therefore, $|\alpha| \leq i - 1$. In this case, $\alpha a^i bp'u'\beta = a^i a^i bp'v'bp'u'$. Then $(\alpha a^i bp'u'\beta)(|\alpha| + i + 1) = b$. On the other hand, $(a^i a^i bp'v'bp'u')(|\alpha| + i + 1) = a$, which is a contradiction.

Case B $p = a^i, a \in X$ and $i \geq 1$.

In this case, $u = a^i u'$ and $v = a^i v'$.

Case B-1 $Int(v') = c \neq a$.

Let $v = a^i cv''$. Since $a^i cv'' a^i u' \in u \triangleright^* u$, $a^i cv'' a^i u' = \alpha(a^i u')\beta$ where $\alpha\beta = a^i u'$. As $Int(u') \neq Int(v') = c$, $|\alpha| \leq i$. If $|\alpha| = 0$, then $u = v$, which is a contradiction. Hence $1 \leq |\alpha| \leq i$ and $\alpha \in a^+$. Consequently, $(\alpha a^i u'\beta)(i+1) = a$. On the other hand, $(a^i cv'' a^i u')(i+1) = c$. This means that $a = c$, which is a contradiction.

Case B-2 $Int(v') = a$.

Let $Int(u') = b$ and let $u = a^i bu''$. Then $a \neq b$. Let $v = a^{i+1}v''$. Since $a^i a^{i+1} v'' bu'' = \alpha(a^i bu'')\beta$ where $\alpha\beta = a^i bu''$. Obviously, $|\alpha| \leq i$. However, in this case, $(\alpha a^i bu''\beta)(|\alpha| + i + 1) = b$ and $(a^i a^{i+1} v'' bu'')(|\alpha| + i + 1) = a$, which is a contradiction.

Consequently, $u = v$.

Conjecture *Let $u \in X^+$. Then $inRes(u \triangleright^* u) = u \triangleright^* u$.*

A word $u \in X^+$ is called an *overlapping word* if $u = \alpha w = w\beta$ for some $\alpha, \beta, w \in X^+$.

Lemma 6.4.2 *Let $u \in X^+$. If $\alpha u = u\beta$ holds for some $\alpha, \beta \in X^+$ with $|\alpha|, |\beta| \leq |u| - 1$, then u is an overlapping word.*

Proof Let $\gamma \in X^+$ be a word such that $u = \gamma\beta$. Then $u = \alpha\gamma$. Consequently, $u = \alpha\gamma = \gamma\beta$. This means that u is an overlapping word.

Proposition 6.4.2 *Let $u \in X^+$. If u is not an overlapping word, then $inRes(u \rhd^* u) = u \rhd^* u$.*

Proof Notice that $u \rhd^* u \subseteq inRes(u \rhd^* u)$. Hence it is enough to show that $v \in u \rhd^* u$ if $v \in inRes(u \rhd^* u)$. Suppose $v \notin u \rhd^* u$. Since $vu \in u \rhd^* u$, there exist $\alpha, \beta, w \in X^+$ such that $|\beta| \leq |u| - 1$, $w \in (v \longrightarrow^* u)$, $wu = \alpha u \beta$ and $\epsilon \in (\alpha\beta \longrightarrow^* u)$. Let $w = \alpha w'$. Then $w' \neq \epsilon$ and $w'u = u\beta$. By Lemma 6.4.2, u is an overlapping word, which is a contradiction. Hence $v \in u \rhd^* u$.

It is unknown whether $inRes(u \rhd^* u) = u \rhd^* u$ holds for any $u \in X^*$.

Chapter 7

Shuffles and Scattered Deletions

In this chapter, we will focus mainly the introducuction of the notion of shuffle residual of a language and discuss its properties and characterizations. The chapter will be organized as follows:

In Section 7.1, the notion of shuffle residual of a language is defined. Some properties of the shuffle residual of a language are described, as well as a characterization of the shuffle residual of a given language. We will also charactarize the shuffle closure of a language.

In Section 7.2, we will provide several conditions for the existence of maximal languages whose shuffle residual equals a given language. A generalization of the notion of shuffle residual of a language will also be introduced.

In Sections 7.3, we will deal with the scattered deletion operation of a language.

The contents of this chapter are based on the results in [41] and [44].

7.1 Shuffle residuals and closures

We will start this section with the definition of the shuffle residual of a language. Let $L \subseteq X^*$ be a language over X. Then the *shuffle residual* of L (denoted by *shRes*(L)) is defined by:

$$shRes(L) = \{x \in X^* \mid \forall u \in L, u \diamond x \subseteq L\}.$$

Example 7.1.1 Let $X = \{a, b\}$. Then we have: (1) *shRes*$(X^*) = X^*$. (2) *shRes*$(L_{ab}) = L_{ab}$. (3) *shRes*$(L) = \{\epsilon\}$ for $L = \{a^n b^n \mid n \geq 0\}$. (4) *shRes*$(L_1) = L_1$ and *shRes*$(L_2) = L_1$ for $L_1 = (a^2)^*$ and $L_2 = aL_1$. (5) *shRes*$(L) = b^*$ for $L = b^*ab^*$. (6) *shRes*$(L) = L$ for $L = aX^*b$.

The following results provide some basic properties of the shuffle residual of a language.

Proposition 7.1.1 *shRes*$(M) \diamond$*shRes*$(N) \subset$ *shRes*$(M \diamond N)$ *and shRes* $(M) \cap$ *shRes*$(N) \subset$ *shRes*$(M \cup N)$.

Proof Let $u \in$ *shRes*$(M) \diamond$ *shRes*(N). This means that there exist $m \in$ *shRes*(M) and $n \in$ *shRes*(N) such that $u \in m \diamond n$. Since the shuffle operation is commutative, we have

$$(M \diamond N) \diamond (m \diamond n) = (M \diamond m) \diamond (N \diamond n) \subseteq M \diamond N.$$

Notice that the equality does not always hold. For example, let $X = \{a, b\}$, $M = ab^*$ and $N = ba^*$. Then *shRes*$(M) = \{\epsilon\} =$ *shRes*(N). On the other hand, $M \diamond N = \{u \in X^* \mid |u|_a \geq 1, |u|_b \geq 1\}$ and hence *shRes*$(M \diamond N) = X^*$.

Now we prove that *shRes*$(M) \cap$ *shRes*$(N) \subseteq$ *shRes*$(M \cup N)$. Let $u \in$ *shRes*$(M) \cap$ *shRes*(N). The fact that $u \in$ *shRes*(M) implies that $M \diamond u \subseteq M$. The fact that $u \in$ *shRes*(N) implies that $N \diamond u \subseteq N$. Consequently we have $(M \cup N) \diamond u \subseteq M \cup N$. Thus $u \in$ *shRes*$(M \cup N)$, i.e. *shRes*$(M \cap N) \subseteq (M \cup N)$.

Notice that the equality does not always hold. For example, if $M = (X^2)^*$ and $N = X(X^2)^*$ then *shRes*$(M) =$ *shRes*$(N) = M$, but *shRes*$(M \cup N) = X^*$.

Proposition 7.1.2 *Let* $L \subseteq X^*$. *Then shRes*(L) *is closed under shuffle. Moreover, shRes*(L) *is a submonoid of* X^*. *If* L *is commutative, then shRes*(L) *is also commutative.*

Proof Let $x, y \in$ *shRes*(L) and $u \in L$. Then $u \diamond x \subseteq L$ and consequently $(u \diamond x) \diamond y \subseteq L$. As shuffle is associative, we have $u \diamond (x \diamond y) \subseteq L$, i.e. $x \diamond y \subseteq shRes(L)$. This implies that $shRes(L)$ is closed under shuffle. In particular, $xy \in x \diamond y$ belongs to $shRes(L)$. Since $\epsilon \in shRes(L)$, $shRes(L)$ is a monoid. Now assume L is commutative. Let $xuvy \in shRes(L)$ for $x, u, v, y \in X^*$. Then $xuvy \diamond w \subseteq L$ for any $w \in L$. As L is commutative, $xvuy \diamond w \subseteq L$ and hence $xvuy \in shRes(L)$. This means that $shRes(L)$ is commutative.

In the following we will provide some properties of $shRes(L)$ and characterize $shRes(L)$ for a given language L. We begin by defining the *iterated shuffle* operation as:

$$L_1 \diamond^* L_2 = \bigcup_{n=0}^{\infty} (L_1 \diamond^n L_2),$$

where $L_1 \diamond^0 L_2 = L_1$ and $L_1 \diamond^{n+1} L_2 = (L_1 \diamond^n L_2) \diamond L_2$.

Lemma 7.1.1 *Let $L \subseteq X^*$ and let $u, v \in shRes(L)$. Then $v \diamond^* u \subseteq shRes(L)$.*

Proof Let $w \in v \diamond^* u$. There exists $k \geq 0$ such that $w \in v \diamond^k u$. We will show, by the induction on k, that $w \in shRes(L)$. If $k = 0$, then $w = v \in shRes(L)$. Assume the assertion holds true for k and take $w \in v \diamond^k u$. Then $w \in \alpha \diamond u$ where $\alpha \in v \diamond^k u$. According to the induction hypothesis, $v \diamond^k u \subseteq shRes(L)$. Therefore, $\alpha \in shRes(L)$. By $\alpha, u \in shRes(L)$ and by Proposition 7.1.2, we have $\alpha \diamond u \subseteq shRes(L)$. This implies $w \in shRes(L)$.

Proposition 7.1.3 *Let $L \subseteq X^*$. Then we have $shRes^2(L) = shRes(shRes(L)) = shRes(L)$.*

Proof Assume $u \in shRes(shRes(L))$. As $\epsilon \in shRes(L)$, we have $u = \epsilon \diamond u \in shRes(L)$, i.e. $shRes(shRes(L)) \subseteq shRes(L)$. Now let $u \in shRes(L)$ and let $v \in shRes(L)$. Obviously, $v \diamond u \in v \diamond^* u$. By Lemma 7.1.1, $v \diamond^* u \subseteq shRes(L)$ and hence $u \in shRes(shRes(L))$, i.e. $shRes(L) \subseteq shRes(shRes(L))$. This completes the proof of the proposition.

We will show that the shuffle residual of a regular langusge is regular.

Lemma 7.1.2 *Let* $L \subseteq X^*$. *Then we have* $shRes(L) \diamond L = L$.

Proof By the definition of $shRes(L)$, $shRes(L) \diamond L \subseteq L$. Since $shRes(L)$ is a monoid, $\epsilon \in shRes(L)$ and hence $L \subseteq shRes(L) \diamond L$. Thus $shRes(L) \diamond L = L$.

Proposition 7.1.4 *If a language* L *is regular, then* $shRes(L)$ *is regular and can be effectively constructed.*

Proof By Lemma 7.1.2, $shRes(L) \diamond L = L$. Therefore, $shRes(L)$ is a solution of the language equation $K \diamond L = L$. From Lemma 4.5.1 and proposition 4.5.1, it follows that $shRes(L)$ is regular and can be effectively constructed.

A language L such that $L \subseteq shRes(L)$ is called *shuffle closed* (or in short, *sh-closed*). A language L is sh-closed if and only if $u \in L$ and $v \in L$ imply $u \diamond v \subseteq L$. As a consequence, every sh-closed language is a submonoid of X^*. Notice that a language L is sh-closed if and only if $L \diamond L \subseteq L$.

In general, submonoids of X^* are not sh-closed. For example, let $X = \{a, b, c\}$ and let $L = (a(bc)^*)^*$. Then L is a submonoid that is not sh-closed, because $a, abc \in L$, but $abac \notin L$.

Proposition 7.1.5 $shRes(L) = L$ *if and only if* L *is sh-closed and* $\epsilon \in L$.

Proof (⇒) Notice that $L \diamond L = shRes(L) \diamond L \subseteq L$. Hence L is sh-closed.

(⇐) If L is sh-closed, then $L \subseteq shRes(L)$. Let $u \in shRes(L)$. Sonce $\epsilon \in L, u = \epsilon \diamond u \in L$. Thus $shRes(L) = L$.

Let $L \subseteq X^*$. Then it is obvious that $L^\diamond$ $(= L \diamond^* L)$ is the smallest sh-closed language containing L. This language is called the *shuffle closure* of L and denoted by $sfc(L)$.

Notice that, if L is a regular (context-free) language, then $sfc(L)$ is not in general a regular (context-free) language. For example, let $X = \{a, b, c\}$ and let $L = \{abc\}$. Then $sfc(L) \cap a^+b^+c^+ = \{a^nb^nc^n \mid n \geq 1\}$ which is not a context-free language though L is finite. Recall that we dealt with shuffle closurse of commutative regular languages.

7.2 Maximal shuffle residuals

In this section, we will provide conditions for the existence of maximal languages whose shuffle residual equals a given language, as well as a generalization of the notion of shuffle residual. Let $L \subseteq X^*$ be an sh-closed language with $\epsilon \in L$. By $\mathcal{M}_X(L)$, we denote the set $\{M \subseteq X^* \mid shRes(M) = L$ and M is maximal in the sense of inclusion relation$\}$.

Definition 7.2.1 An sh-closed language $L \subseteq X^*$ is said to be an *RSS-type language* if it contains a regular ssh-closed language L_0 with $alph(L) = alph(L_0)$.

Proposition 7.2.1 *An sh-closed language $L \subseteq X^*$ is an RSS-type language if and only if, for any $a \in alph(L), a^+ \cap L \neq \emptyset$.*

Proof ($\Rightarrow$) Let $a \in alph(L)$. Then $a \in alph(L_0)$. Since L_0 is an ssh-closed language, by Proposition 5.2.4, $a^+ \cap L_0 \neq \emptyset$ and hence $a^+ \cap L \neq \emptyset$.

($\Leftarrow$) Let $alph(L) = \{a_1, a_2, ..., a_n\}$ and let p_i be a positive integer such that $a_i^{p_i} \in L$ for any $i, 1 \leq i \leq n$. Moreover, let $L_0 = (a_1^{p_1})^* \diamond (a_2^{p_2})^* \diamond ... \diamond (a_n^{p_n})^*$. Then $L_0 \subseteq L$ and L_0 is a regular ssh-closed language.

In what follows, $L \subseteq X^*$ is assumed to be an RSS-type language containing a regular ssh-closed language L_0 with $alph(L) = alph(L_0)$.

Lemma 7.2.1 *Let $M \subseteq X^*$ with $shRes(M) = L$. Then M can be represented as $M = \bigcup_{i \in I}(\alpha_i \diamond L_0)$ where $\alpha_i \in X^*, i \in I$ and I is some index set.*

Proof Obvious from the fact $L_0 \diamond M = M$.

Lemma 7.2.2 *Let $M \subseteq X^*$ with $shRes(M) = L$. If $alph(L) = X$, then there exists a positive integer p satisfying the following condition: For any $u \in X^*$, there exists $\beta \in X^*$ such that $|\beta| \leq p$ and $u \in \beta \diamond L_0$.*

Proof Let $X = \{a_1, a_2, ..., a_n\}$. Since L_0 is ssh-closed, for any $i = 1, 2, \ldots, n$, there exists a positive integer p_i such that $(a_i^{p_i})^* \subseteq L_0$. Now let $u \in X^*$. Then $u \in u' \diamond (a_1^{p_1})^* \diamond (a_2^{p_2})^* \diamond \ldots \diamond (a_n^{p_n})^*$ where $0 \leq |u'|_{a_i} < p_i$ for any $i = 1, 2, \ldots, n$. Let $p = \sum_{i=1}^{n}(p_i - 1)$. Then $u \in u' \diamond L_0$ and $|u'| \leq p$.

Definition 7.2.2 By C_β, we denote the set $\beta \diamond L_0$.

Lemma 7.2.3 *Let $u, v \in C_\beta$ and let $u \leq_h v$. Then $v \diamond L_0 \subseteq u \diamond L_0$.*

Proof First, $v \mapsto u \neq \emptyset$. Since $u, v \in C_\beta$ and L_0 is commutative, we have $v \mapsto u \subseteq L_0$. Hence $v \in u \diamond L_0$. Therefore, $v \diamond L_0 \subseteq (u \diamond L_0) \diamond L_0 = u \diamond L_0$.

Proposition 7.2.2 *Assume that $alph(L) = X$. Let $M \subseteq X^*$ with $shRes(M) = L$. Then M is regular.*

Proof Let $M = \bigcup_{i \in I}(\alpha_i \diamond L_0)$. Then by Lemma 7.2.2, there exists a positive integer q and $\beta_j \in X^*, j = 1, 2, \ldots, q$ such that $\{\alpha_i \mid i \in I\} = \bigcup_{j \in \{1,2,\ldots,q\}}(\beta_j \diamond L_0)$. Let $D_j = C_{\beta_j} \cap \{\alpha_i \mid i \in I\}$ for any $j = 1, 2, \ldots, q$. Notice that each D_j contains a maximal hypercode H_j in D_j. Let $H_j^{\downarrow} = \{u \in D_j \mid \exists v \in H_j, u \leq_h v\}$ for any $j = 1, 2, \ldots, q$ and let $E = \bigcup_{j \in \{1,\ldots q\}} H_j^{\downarrow}$. Notice that $E \subseteq \{\alpha_i \mid i \in I\}$ and E is finite. Let $\alpha \in \{\alpha_i \mid i \in I\}$. Then there exists $j = 1, 2, \ldots, q$ such that $\alpha \in D_j$. By the definition of $H_j^{\downarrow}$, $\alpha \in H_j$ or $\alpha_k \leq_h \alpha$ for some $\alpha_k \in H_j^{\downarrow}$. In the former case, $\alpha \diamond L_0 \subseteq E \diamond L_0$. In the latter case, $\alpha \in \alpha_k \diamond L_0$ and hence $\alpha \diamond L_0 \subseteq \alpha_k \diamond L_0 \subseteq E \diamond L_0$. Therefore, $E \diamond L_0 \subseteq \bigcup_{i \in I}(\alpha_i \diamond L_0) \subseteq E \diamond L_0$ and $M = E \diamond L_0$. Since E and L_0 are regular, M is regular.

Corollary 7.2.1 *An RSS-type language is regular.*

Proof Since $L = shRes(M)$ and M is regular, L is regular.

Notice that, if $alph(L) \subset X$, then the statement in Proposition 7.2.2 does not hold true. For instance, let $M = L \cup (\bigcup_{n \geq 1}(b^{n!} \diamond L))$ where L is an RSS-type language and $b \in X \setminus alph(L)$. Then $shRes(M) = L$ but M is not regular.

Theorem 7.2.1 *Assume $alph(L) = X$. Let $M \subseteq X^*$ with $shRes(M) = L$. Then there exists $N \supseteq M$ such that $N \in \mathcal{M}_X(L)$.*

Proof Let $M = M_0 \subset M_1 \subset M_2 \subset \cdots$ be an ascending chain of languages such that $shRes(M_i) = L$ for any $i \geq 0$. Moreover, let $M_i = \bigcup_{j \in I_i}(\alpha_{ij} \diamond L_0)$ for any $i \geq 0$. From the same reason as in the proof of Proposition 7.2.2, it follows that $\{\alpha_{ij} \mid i \geq 0, j \in I_i\} \subseteq \bigcup_{j \in \{1,2,\ldots,q\}} C_{\beta_j}$. Let $D_k = C_{\beta_k} \cap \{\alpha_{ij} \mid i \geq 0, j \in I_i\}$ for any $k = 1, 2, \ldots, q$. Let H_k be a maximal hypercode in D_k, let $H_k^{\downarrow} = \{u \in D_k \mid \exists v \in H_k, u \leq_h v\}$ for any $k = 1, 2, \ldots, q$ and let $E = \bigcup_{k \in \{1,2,\ldots,q\}} H_k^{\downarrow}$. Then E is finite. Suppose $M = M_0 \subset M_1 \subset M_2 \subset \cdots$ is an infinite ascending chain. Since E is finite, there exists a possitive integer r such that $E \diamond L_0 \subseteq M_r$. Let $\alpha_{(r+1)t} \in M_{r+1} \setminus M_r$. In the same way as in the proof of Proposition 7.2.2, $\alpha_{(r+1)t} \diamond L_0 \subseteq E \diamond L_0 \subseteq M_r$, which is a contradiction. Hence $M = M_0 \subset M_1 \subset M_2 \subset \cdots$ is always a finite ascending chain. Consequently, $N = M_r \in \mathcal{M}_X(L)$.

The situation is completely different for the case $alph(L) \subset X$. Let $alph(L) = Y \subset X$ and let $Z = X \setminus Y$. Now let $Z^* = \{z_0, z_1, z_2, \ldots\}$ where $z_0 = \epsilon$ and $|z_i| \leq |z_{i+1}|$ for any $i \geq 0$. Now suppose there exists $M \in \mathcal{M}_X(L)$ and $M = \bigcup_{i \geq 0}(z_i \diamond M_i)$ where $M_i \subseteq Y^*$ for any $i \geq 0$. Notice that $shRes(M_i) \supseteq L$ for any $i \geq 0$.

Lemma 7.2.4 *$M_i \neq \emptyset$ for any $i \geq 0$.*

Proof Suppose $M_0 = \emptyset$. Consider $N = M \cup L$. It is obvious that $N \supset M$ and $shRes(N) \supseteq L$. Since $M \in \mathcal{M}_X(L)$, $shRes(N) \supset L$. Let $x \in shRes(N) \setminus L$. Then there exists $m_i \in z_i \diamond M_i, i \geq 1$ such that $(m_i \diamond x) \cap L \neq \emptyset$. However, this is impossible because, for any $u \in L, |u|_Z = 0$ but, for any $v \in m_i \diamond x, |v|_Z \geq |m_i|_Z \geq 1$. Hence $M_0 \neq \emptyset$. Now suppose $M_i = \emptyset$ for some $i \geq 1$. Consider $N = M \cup (z_i \diamond L)$. Obviously, $N \supset M$ and $shRes(N) \supseteq L$. By the maximality of M, $shRes(N) \supset L$. Hence there exists $x \in shRes(N) \setminus L$ such that $(x \diamond M) \cap (z_i \diamond L) \neq \emptyset$. This implies that $|x|_Z > 0$. Now consider $x^{|z_i|+1} \in shRes(N) \setminus L$. Then there exists $m \in M$ such that $(x^{|z_i|+1} \diamond m) \cap (z_i \diamond L) \neq \emptyset$. This yields a contradiction because, for any $u \in z_i \diamond L, |u|_Z = |z_i|$ but, for any $v \in x^{|z_i|+1} \diamond m, |v|_Z \geq |z_i| + 1$. Hence $M_i \neq \emptyset$ for any $i \geq 0$.

Lemma 7.2.5 *Let* $\boldsymbol{N} = \{i \mid i \geq 0, M_i \neq Y^*\}$. *Then* $\boldsymbol{N}$ *is infinite.*

Proof Suppose there exists a positive integer n_0 such that, for any $n \geq n_0, M_n = Y^*$. Consider $z_n^2 \in Z^+$. Obviously, $z_n^2 \diamond M \subseteq M$. Hence $z_n^2 \in shRes(M)$, which is a contradiction. This completes the proof of the lemma.

Now let $K_i = shRes(M_i)$ for any $i \geq 0$. Recall that $K_i \supseteq L$ for any $i \geq 0$.

Lemma 7.2.6 $L = \bigcap_{i\geq 0} K_i$.

Proof Obviously, $L \subseteq \bigcap_{i\geq 0} K_i$. Let $x \in \bigcap_{i\geq 0} K_i$. Since $x \in K_0, x \diamond M_0 \subseteq M_0$. Therefore, $|x|_Z = 0$ and $(z_i \diamond M_i) \diamond x \subseteq (z_i \diamond M_i)$ for any $i \geq 1$. This implies that $x \diamond M \subseteq M$, i.e. $x \in L$. This completes the proof of the lemma.

Lemma 7.2.7 *Let* $N = (\bigcup_{i\geq 0, i\neq k}(z_i \diamond M_i)) \cup (z_k \diamond Y^*)$. *Then* $shRes(N) = \bigcap_{i\geq 0, i\neq k} K_i$.

Proof That $\bigcap_{i\geq 0, i\neq k} K_i \subseteq shRes(N)$ is obvious. Let $x \in shRes(N)$. If $|x|_Z > 0$, then $x^{|z_k|+1} \in shRes(N)$. Since $x^{|z_k|+1} \diamond N \subseteq \bigcup_{i\geq k+1}(z_i \diamond M_i)$ and $z_k \diamond M_k \subseteq z_k \diamond Y^*, x^{|z_k|+1} \diamond M \subseteq M$, i.e. $x^{|z_k|+1} \in shRes(M) = L$, which is a contradiction. Hence $|x|_Z = 0$. Since $|x|_Z = 0, x \diamond M_i \subseteq M_i$ for any $i, k \neq i \geq 0$, i.e. $x \in \bigcap_{i\geq 0, i\neq k} K_i$. This completes the proof of the lemma.

Let $A = \{w \in X^* \mid \exists i \geq 0, w \notin K_i, \forall j \neq i, w \in K_j\}$. Notice that $A = \{w \in X^* \mid \exists i \geq 0, w \notin K_i, \forall j \neq i, w \diamond L_0 \subseteq K_j\}$. As the proof of Proposition 7.2.2, there exists a finite set B with $B \subseteq A$ such that, for any $w \in A$ there exists $w' \in B$ with $w \in w' \diamond L_0$. Now let $w \notin K_i$ and let $w' \notin K_j$. Suppose $i \neq j$. Then $w' \in K_i$ and hence $w \in w' \diamond L_0 \subseteq K_i$, which is a contradiction. Therefore, $i = j$. Let $\boldsymbol{B} = \{w_1, w_2, ..., w_r\}$. Moreover, for any $i = 1, 2, \ldots, r$, we choose some integer $f(i)$ such that $w_i \notin K_{f(i)}$. Then the following is now obvious.

Lemma 7.2.8 *Let* $w \in A$. *Then* $w \notin \bigcap_{1\leq i\leq r} K_{f(i)}$.

Proposition 7.2.3 *Let $alph(L) = Y \subset X$. Then $\mathcal{M}_X(L) = \emptyset$.*

Proof By Lemma 7.2.5, there exists a positive integer t such that $M_t \neq Y^*$ and $t \notin \{f(1), f(2), \ldots, f(r)\}$. Let $u \notin L$. If $u \in \boldsymbol{A}$, then $u \notin \bigcap_{1\leq i\leq r} K_{f(i)}$ and hence $u \notin \bigcap_{i\geq 0, i\neq t} K_i$. If $u \notin \boldsymbol{A}$, then there exist at least two distinct integers i and j such that $u \notin K_i \cup K_j$. Therefore, $u \notin \bigcap_{i\geq 0, i\neq t} K_i$. Now let $N = (\bigcup H_{i\geq 0, i\neq t}(z_i \diamond M_i)) \cup (z_t \diamond Y^*)$. Obviously, $N \supset M$. By Lemma 7.2.7, *shRes*$(N) = \bigcap_{i\geq 0, i\neq t} K_i = L$. This contradicts the maximality of M and hence $\mathcal{M}_X(L) = \emptyset$.

We consider now similar questions for a generalization of the notion of shuffle residual. The shuffle residual of a language consists of the words x whose shuffle $L \diamond x$ is completely included in L. We can relax this condition by only requiring that at least one word from $L \diamond x$ belongs to L. The notion obtained in this way generalizes the notion of shuffle residual. More precisely, the *generalized shuffle residual* of a language L, denoted by *g-shRes*(L) is defined as follows.

Definition 7.2.3 Let $M \subseteq X^*$. Then *g-shRes*$(M) = \{x \in X^* \mid \exists y \in M, (x \diamond y) \cap M \neq \emptyset\}$.

The following results give some properties of the generalized shuffle residual of a language.

Proposition 7.2.4 *If M is a semigroup, then the generalized shuffle residual of M is a monoid.*

Proof It is obvious that $\epsilon \in M$. Let $x, y \in$ *g-shRes*(M). Then there exist $z_1, z_2 \in M$ such that $(x \diamond z_1) \cap M \neq \emptyset$ and $(y \diamond z_2) \cap M \neq \emptyset$. This implies that $(xy \diamond z_1 z_2) \cap M \neq \emptyset$, i.e. $xy \in$ *g-shRes*(M).

Proposition 7.2.5 *If M is sh-closed, then g-shRes(M) is sh-closed.*

Proof If $x, y \in$ *g-shRes*(M) this means that there exist $z_1, z_2 \in M$ such that $(x \diamond z_1) \cap M \neq \emptyset$ and $(y \diamond z_2) \cap M \neq \emptyset$.

Let z be a word in $x \diamond y$. As $z_1 \diamond z_2 \subseteq M$ we have

$$z \diamond (z_1 \diamond z_2) \subseteq (x \diamond y) \diamond (z_1 \diamond z_2) = (x \diamond z_1) \diamond (y \diamond z_2),$$

which implies that $(z \diamond (z_1 \diamond z_2)) \cap M \neq \emptyset$.

Proposition 7.2.6 *If $M \subseteq X^*$ is regular, then g-shRes(M) is regular.*

Proof Let M be a regular language accepted by a finite acceptor $\mathbf{A} = (S, X, \delta, s_0, F)$. Denote by $\bar{X} = \{\bar{a} | \ a \in X\}$ and $\tilde{X} = X \cup \bar{X}$.

Consider the function $\delta_1 : S \times \tilde{X} \longrightarrow S$ defined by $\delta_1(s, a) = \delta(s, a)$ if $a \in X$ and $\delta_1(s, \bar{a}) = s$ if $\bar{a} \in \bar{X}$.

Consider another function $\delta_2 : S \times \tilde{X} \longrightarrow S$ defined by $\delta_2(s, a) = \delta(s, a)$ if $a \in X$ and $\delta_2(s, \bar{a}) = \delta(s, a)$ if $\bar{a} \in \bar{X}$.

Define now the acceptor

$$\tilde{A} = (S, \tilde{X}, \tilde{\delta}, (s_0, s_0), \{(s, t) | \ s \in F, t \in F\})$$

where the transition function $\tilde{\delta} : (S \times S) \times \tilde{X} \longrightarrow S \times S$ is defined by $\tilde{\delta}((s, t), b) = (\delta_1(s, b), \delta_2(t, b))$ for $b \in \tilde{X}$.

Then we have

$$\begin{aligned} L(\tilde{A}) = \ & \{z \in \tilde{X}^* \mid \exists x_i \in \{\epsilon\} \cup X, \exists y_i \in \{\epsilon\} \cup X, \\ & z = x_1\bar{y_1}x_2\bar{y_2} \cdots x_n\bar{y_n}, x_1x_2 \cdots x_n \in M, \\ & x_1y_1x_2y_2 \cdots x_ny_n \in M, \ \text{where } \bar{\epsilon} = \epsilon\}. \end{aligned}$$

Let ρ be the homorphism of $\tilde{X}^*$ into X^* defined by $\rho(a) = \epsilon$ if $a \in X$ and $\rho(\bar{a}) = a$ if $\bar{a} \in \bar{X}$.

Then it can be seen that *g-shRes*$(M) = \rho(L(\tilde{A}))$ and hence *g-shRes*(M) is regular.

By $\mathcal{G}$, we denote the set $\{L \subseteq X^* \mid \exists M \subseteq X^*$ with *g-shRes*$(M) = L\}$.

Remark 7.2.1 The following statement is not always true: For any $L \in \mathcal{G}$ there exists a maximal $M \subseteq X^*$ such that *g-shRes*$(M) = L$.

Proof $\{\epsilon\} \in \mathcal{G}$. Suppose that there exists a maximal $M \subseteq X^*$ such that *g-shRes*$(M) = \{\epsilon\}$. Let $\alpha \notin M$ and let $\tilde{M} = M \cup \{\alpha\}$. It is obvious that $\{\epsilon\} \in$ *g-shRes*$(\tilde{M})$. Suppose that $\epsilon \neq x \notin$ *g-shRes*$(\tilde{M})$. Since *g-shRes*$(\tilde{M})$ is a monoid, $x^n \in$ *g-shRes*$(\tilde{M})$ for any $n, n \geq 1$. Let $n \geq 1$ such that $n|x| > |\alpha|$. Notice that $(M \cup \{\alpha\}) \diamond x^n \subseteq M \cup \alpha$. However, since $x^n \notin$*g-shRes*(M), there exists $m \in M$ such that $x^n \diamond m \nsubseteq M$. Hence $\alpha \in x^n \diamond m$, but this contradicts the assumption $n|x| > |\alpha|$. Therefore *g-shRes*$(\tilde{M}) = \{\epsilon\}$. This means that M is not maximal.

7.3 Scattered deletion closure

First, we will define the scattered deletions of words and languages.

Definition 7.3.1 Let $u, v \in X^*$. Then $u \mapsto v = \{u_1 u_2 \cdots u_r \mid u = u_1 v_1 u_2 v_2 \cdots u_r v_r, v = v_1 v_2 \cdots v_r, r \geq 1, u_1, v_1, \ldots, u_r, v_r \in X^*\}$ is called the *scattered deletion* of v from u. For languages $L_1, L_2 \subseteq X^*$, the language $\bigcup_{u \in L_1, v \in L_2} u \mapsto v$ is called the *scattered deletion* of L_2 from L_1.

The following fundamental result is due to [54].

Proposition 7.3.1 *The scattered deletion of a regular language from a regular language is regular.*

Proof Let $L, K \subseteq X^*$ be regular languages. Let $\boldsymbol{A} = (S, X, \delta, s_0, F)$ and $\boldsymbol{B} = (T, X, \gamma, t_0, G)$ be finite acceptors such that $\mathcal{T}(\boldsymbol{A}) = L$ and $\mathcal{T}(\boldsymbol{B}) = K$, respectively. Moreover, let $X' = \{a' \mid a \in X\}$.

Consider the following finite acceptors $\boldsymbol{A}' = (S, X \cup X', \delta', s_0, F)$ and $\boldsymbol{B}' = (T, X \cup X', \gamma', t_0, G)$: (1) $sa^{A'} = sa'^{A'} = sa^A$ for any $s \in S$ and any $a \in X$. (2) $ta^{B'} = t$ for any $t \in T$ and any $a \in X$, and $ta'^{B'} = ta^B$ for any $t \in T$ and any $a' \in X'$.

Let ρ be the following homomorphism of $(X \cup X')^*$ onto X^*: $\rho(a) = a$ and $\rho(a') = \epsilon$ for any $a \in X$. By the similar way in the proof of Proposition 6.2.1, it can be verified that $\rho(\mathcal{T}(\boldsymbol{A}') \cap \mathcal{T}(\boldsymbol{B}') \cap (X \cup X')^*) = L \longrightarrow K$. Since $\mathcal{T}(\boldsymbol{A}') \cap \mathcal{T}(\boldsymbol{B}') \cap (X \cup X')^*$ is regular, $L \mapsto K$ is regular.

As in the case of deletion, the above proposition does not hold for context-free languages. We can take the same example, i.e. let $X = \{a, b, c, d\}$, let $L = \{a^n b^n c^m d^m \mid n, m \geq 1\}$ and let $K = \{b^k c^k \mid k \geq 1\}$. Then $L \mapsto K = \{a^n b^{n-k} c^{m-k} d^m \mid n, m \geq k \geq 1\}$ is not context-free.

Let $L \subseteq X^*$. Then the set of *sparse* subwords $sps(L)$ of L is defined as follows: $sps(L) = \{u \in X^* \mid u = a_1 \cdots a_k$ and there exist $v_i \in X^*, i = 1, 2, \ldots, k+1$ such that $v_1 a_1 v_2 a_2 \cdots v_k a_k v_{k+1} \in L\}$.

We also define the *scattered deletion residual* $sdRes(L)$ of L as follows: $sdRes(L) = \{x \in sps(L) \mid \forall u \in L, u \mapsto x \subseteq L\}$.

In the above, the condition that $x \in sps(L)$ has been added because otherwise $sdRes(L)$ would contain irrelevant words which are not sparse subwords of any word of L and thus yield $\emptyset$ as a result of the scattered deletion from L.

Example 7.3.1 Let $X = \{a, b\}$. Then (1) $sdRes(X^*) = X^*$. (2) $sdRes(L_{ab}) = L_{ab}$ where $L_{ab} = \{u \in X^* \mid |u|_a = |u|_b\}$. (3) For $L = \{a^n b^n \mid n \geq 0\}$, $sdRes(L) = L$. (4) For $L = b^*ab^*$, $sdRes(L) = b^*$.

The following proposition gives some basic properties of the scattered deletion residual of a language.

Proposition 7.3.2 *Let $L \subseteq X^*$. (i) If $x, y \in sdRes(L)$ and $xy \in sps(L)$, then $xy \in sdRes(L)$. (ii) If $sps(L)$ is a submonoid of X^*, then $sdRes(L)$ is a submonoid of X^*. (iii) If L is a commutative language, then $sdRes(L)$ is also commutative.*

Proof (i) Let $x, y \in sdRes(L)$ with $xy \in sps(L)$. If $u \in L$, then $u = u_1x_1u_2x_2 \cdots u_kx_ku_{k+1}y_1u_{k+2}y_2 \cdots u_{n+k}y_nu_{n+k+1}$. Since $x \in sdRes(L)$, we have $u_1 \cdots u_{k+1}y_1 \cdots y_nu_{n+k+1} \in L$. Moreover, since $y \in sdRes(L)$, we can conclude that $u_1 \cdots u_k \cdots u_{n+k+1} \in L$. As the initial decomposition of u was arbitrary, we deduce that $u \mapsto xy \subseteq L$, which implies $xy \in sdRes(L)$.

(ii) Immediate from (i).

(iii) Let $x = x_1x_2 \cdots x_k \in sdRes(L)$, $x_i \in X^*$ for $1 \leq i \leq k$. As $x \in sps(L)$ and L is commutative, $\Psi^{-1}\Psi(x) \subseteq sps(L)$ where Ψ is the Parikh mapping. Let $y = y_1y_2 \cdots y_k$, $y_i \in X$, $i \geq 0$, be a word in $\Psi^{-1}\Psi(x)$ and let $u = u_1y_1 \cdots u_ky_ku_{k+1} \in L$, $u_i \in X^*$, $y_i \in X$. As L is commutative, the word $u_1x_1 \cdots u_kx_ku_{k+1}$ is in L and, as $x \in sdRes(L)$, we have that $u_1 \cdots u_ku_{k+1} \in L$. This implies $u \mapsto y \subseteq L$, which means $y \in sdRes(L)$.

In the following, we construct the set $sdRes(L)$ for a given language L.

Proposition 7.3.3 *If L is a language in X^* then $sdRes(L) = (L \mapsto L^c)^c \cap sps(L)$ where K^c means $X^* \setminus K$ for $K \subseteq X^*$.*

Proof Let $x \in sdRes(L)$. From the definition of $sdRes(L)$ it follows that $x \in sps(L)$. Assume that $x \notin (L \mapsto L^c)^c$. This means there exists $w \in L$, $v \in L^c$ such that $x \in (w \mapsto v)$. This implies $v \in (w \mapsto x)$. We arrived at a contradiction as $x \in sdRes(L)$ but there exists a word $w \in L$ with $(w \mapsto x) \cap L^c = v \neq \emptyset$.

For the other inclusion, let $x \in (L \mapsto L^c)^c \cap sps(L)$. As $x \in sps(L)$, if $x \notin sdRes(L)$ then there exists $w \in L$ such that $v \in (w \mapsto x) \cap L^c \neq \emptyset$. This implies $x \in (w \mapsto v) \subseteq (L \mapsto L^c)$, which is a contradiction with the initial assumption about x.

The following result connects the notions of shuffle and scattered deletion.

Proposition 7.3.4 *Let $L \subseteq X^*$ be an sh-closed language. Then L is sd-closed if and only if $L = (L \mapsto L)$.*

Proof If L is sd-closed, $L \mapsto L \subseteq L$. Now let $u \in L$. Since L is sh-closed, $uu \in L$. Therefore $u \in (L \mapsto L)$, i.e. $L \subseteq (L \mapsto L)$. We can conclude that $L = (L \mapsto L)$. The other implication is obvious.

If L is a nonempty language and if D_L is the family of all sd-closed languages L_i containing L, then the intersection

$$\bigcap_{L_i \in D_L} L_i$$

of all sd-closed languages containing L is an sd-closed language called the *scattered deletion closure* of L, or in short, *sd-closure* of L. The sd-closure of L is the smallest sd-closed language containing L.

We will now define a sequences of languages whose union is the sd-closure of a given language L. Let:

$$sdc_0(L) = L$$

$$sdc_1(L) = sdc_0(L) \mapsto (sdc_0(L) \cup \{\epsilon\})$$

$$sdc_2(L) = sdc_1(L) \mapsto (sdc_1(L) \cup \{\epsilon\})$$

$$\cdots$$

$$sdc_{k+1}(L) = sdc_k(L) \mapsto (sdc_k(L) \cup \{\epsilon\})$$

$$\ldots$$

Clearly $sdc_k(L) \subseteq sdc_{k+1}(L)$. Let $sdc(L) = \bigcup_{k\geq 0} sdc_k(L)$.

Proposition 7.3.5 *$sdc(L)$ is the sd-closure of the language L.*

Proof Clearly $L \subseteq sdc(L)$. Let now $v \in sdc(L)$ and $u \in sdc(L)$. Then $v \in sdc_i(L)$ and $u \in sdc_j(L)$ for some integers $i, j \geq 0$. If $k = max\{i, j\}$, then $v \in sdc_k(L)$ and $u \in sdc_k(L)$. This implies $(u \mapsto v) \subseteq sdc_{k+1}(L) \subseteq sdc(L)$. Therefore $sdc(L)$ is an sd-closed language containing L.

Let T be an sd-closed language such that $L = sdc_0(L) \subseteq T$. Since T is sd-closed, if $sdc_k(L) \subseteq T$ then $sdc_{k+1}(L) \subseteq T$. By an induction argument, it follows that $sdc(L) \subseteq T$. This completes the proof of the proposition.

It can be proved that the languages $sdc_k(L)$, $k \geq 0$, is regular if $L \subseteq X^*$ is regular.

By Proposition 7.3.1, the family of regular languages is closed under scattered deletion. Therefore, if L is regular, then the languages $sdc_k(L)$, $k \geq 0$, are also regular. However, this result does not imply that $sdc(L)$ is regular. Proposition 7.3.6 and Proposition 7.3.7 show different cases.

Recall that, for a language L, the principal congruence P_L of L is defined by: $u \equiv v(P_L)$ if and only if $xuy \in L \Leftrightarrow xvy \in L$ for any $x, y \in X^*$.

When the principal congruence of L has a finite index (finite number of classes) the language L is regular.

If L is commutative, we have the following result.

Proposition 7.3.6 *Let $L \subseteq X^*$ be a regular language. If L is commutative, then its scattered deletion closure $sdc(L)$ is commutative and regular.*

Proof Let us prove first that $sdc(L)$ is commutative. To this end, it is sufficient to show that $sdc_{k+1}(L)$ is commutative if $sdc_k(L)$ is commutative. Let $xuvy \in sdc_{k+1}(L)$. If $xuvy \in sdc_k(L)$, then we are done. Otherwise, by the definition of $sdc_{k+1}(L)$, there exist $w, z \in sdc_k(L)$ such that $w \in (xuvy \diamond z)$. Since $sdc_k(L)$ is commutative, $xuvyz \in sdc_k(L)$ and $xvuyz \in sdc_k(L)$. From the fact that $z, xvuyz \in sdc_k(L)$ and the definition of $sdc_{k+1}(L)$, it follows that $xvuy \in sdc_{k+1}(L)$, i.e. $sdc_{k+1}(L)$ is commutative.

We will show next that $sdc(L)$ is regular. To this end, we show that if $u \equiv v(P_{sdc_k(L)})$ then $u \equiv v(P_{sdc_{k+1}(L)})$. Let $u \equiv v(P_{sdc_k(L)})$ and let $xuy \in sdc_{k+1}(L)$. By the definition of $sdc_{k+1}(L)$, there exists $w, z \in sdc_k(L)$ such that $w \in (xuy \diamond z)$. Since $sdc_k(L)$ is commutative, $xuyz \in sdc_k(L)$. Hence $xvyz \in sdc_k(L)$. From the fact that $z \in sdc_k(L)$ and by the definition of $sdc_{k+1}(L)$, it follows that $xvy \in sdc_{k+1}(L)$. In the same way, $xvy \in sdc_{k+1}(L)$ implies $xuy \in sdc_{k+1}(L)$. Consequently, $u \equiv v(P_{sdc_{k+1}(L)})$ holds. This means that the number of congruence classes of $P_{sdc_{k+1}(L)}$ is smaller or equal to that of $P_{sdc_k(L)}$. Notice that

$$sdc_0(L) \subseteq sdc_1(L) \subseteq \cdots \subseteq sdc_n(L) \subseteq sdc_{n+1}(L) \subseteq \cdots.$$

Hence there exist positive integers k, n_0 such that the number of congruence classes of $P_{sdc_n(L)}$ equals k for any $n \geq n_0$. Since the number of languages over X whose number of congruence classes is constant is finite, we have $sdc_t(L) = sdc_{t+1}(L)$ for some $t \geq 1$. Thus $sdc(L) = sdc_t(L)$ which implies that $sdc(L)$ is regular (see also [43]).

However, in general the above situation does not happen. The proof is due to [44].

Proposition 7.3.7 *If $|X| \geq 2$, then there is a regular (context-free) language $L \subseteq X^*$ such that $sdc(L)$ is not regular (context-free).*

Proof Let $X = \{a, b, \ldots\}$ and let $L = (ab)^+(abba)^+(babbaa)^+ \cup \{bbaa\}$. Then it is easy to see that $|u|_a = |u|_b$ for any $u \in sdc(L)$. Since $a^n(aba)^n(babb)^n \in ((ab)^+(abba)^+(babbaa)^+ \mapsto^* bbaa)$, $a^n(aba)^n (babb)^n \in sdc(L) \cap a^+(aba)^+(babb)^+$ for any $n, n \geq 1$. In the above, $\mapsto^*$ means a consecutive application of $\mapsto$. Suppose $sdc(L)$ is context-free. Then $sdc(L) \cap a^+(aba)^+(babb)^+$ is context-free. Since $sdc(L) \cap$

$a^+(aba)^+(babb)^+$ is context-free, we have $a^n(aba)^n(babb)^n = uvwxy$ where $1 \leq |vx| \leq n$ and $uv^iwx^iy \in sdc(L) \cap a^+(aba)^+(babb)^+$ for any $i \geq 1$ for a sufficiently large number n. However, in this case, $|uv^iwx^iy|_a \neq |uv^iwx^iy|_b$ for any $i \geq 2$, which is a contradiction. Hence $sdc(L)$ is not context-free though L is regular. This completes the proof of the proposition.

Chapter 8

Directable Automata

An automaton $\boldsymbol{A} = (S, X, \delta)$ is called a *directable automaton* if there exists $w \in X^*$ such that $sw^A = tw^A$ for any $s, t \in S$. Then a word w is called a *directing word* of the directable automaton $\boldsymbol{A}$.

Fact *Let $\boldsymbol{A} = (S, X, \delta)$ be an automaton. Then $\boldsymbol{A}$ is directable if and only if for any $s, t \in S$, there exists $u \in X^*$ such that $su^A = tu^A$.*

Proof ($\Rightarrow$) Obvious.

($\Leftarrow$) Let T be a maximal subset of S such that $\exists w \in X^*, \forall t, t' \in T, tw^A = t'w^A$. If $T = S$, then $\boldsymbol{A}$ is a directable automaton. Suppose $T \neq S$. Then there exists $s \in S \setminus T$ such that $sw^A \neq tw^A$ for any $t \in T$. Moreover, there there exists $w' \in X^*$ such that for any $t \in T, (sw^A)w'^A = (tw^A)w'^A$, i.e. $s(ww')^A = t(ww')^A$. Let $T' = T \cup \{s\}$. Then $t_1(ww')^A = t_2(ww')^A$ for any $t_1, t_2 \in T'$. This contradicts the maximality of T. Hence $\boldsymbol{A}$ is directable.

Recall that, in Chapter 3, we considered the class of directable (strongly directable) automata. In this chapter, we will deal with directable automata and then nondeterministic directable automata. For the case of nondeterministic automata, the notion of directability is not uniquely determined. We will introduce three kinds of directabilities for nondeterministic automata. The contents of this chapter consist of the results in [25], [26] and [28].

8.1 Deterministic case

We will start this section by proving that the set of directing words of a directable automaton is a regular language.

Proposition 8.1.1 *Assume that* $\boldsymbol{A} = (S, X, \delta)$ *is a directable automata. Then the set of directing words* $\mathrm{D}(\boldsymbol{A})$ *of* $\boldsymbol{A}$ *is a regular language.*

Proof Let $\boldsymbol{A} = (S, X, \delta)$ be a directable automata. For any $a \in X$ and any subset T of S, we define $\overline{\delta}(T, a) = \bigcup_{t \in T}\{\delta(t, a)\}$.

Then we have a finite acceptor $\overline{\boldsymbol{A}} = (\{T \mid T \subseteq S\}, X, \overline{\delta}, S, \{\{s\} \mid s \in S\})$. Since $\boldsymbol{A}$ is directable, we have $\mathcal{T}(\overline{\boldsymbol{A}}) = \mathrm{D}(\boldsymbol{A})$. Hence $\mathrm{D}(\boldsymbol{A})$ is regular.

Let $\boldsymbol{A} = (S, X, \delta)$ be a directable automaton. By $d(\boldsymbol{A})$, we denote the value $min\{|w| \mid w \in \mathrm{D}(\boldsymbol{A})\}$. Moreover, $d(n)$ denotes the value $max\{d(\boldsymbol{A}) \mid \boldsymbol{A} = (S, X, \delta)$ is a directable automaton with n states$\}$. In the definition of $d(n)$, X ranges over all finite nonempty alphabets.

In [11], Černý conjectured the following.

Conjecture *For any* $n \geq 1, d(n) = (n-1)^2$.

Calculating all possible cases, Černý confirmed that the above conjecture holds true up to $n = 5$. However, the above is still an open problem for the general case. We will only show that $d(n)$ is the order of $\mathcal{O}(n^3)$. Better estimations can be found in [58] and [59]. First, we will introduce the following result by Černý that suggests the reason why Černý reached the above conjecture.

Proposition 8.1.2 *For any* $n \geq 1$, *we have* $(n-1)^2 \leq d(n)$.

Proof Consider the following automaton:

$\boldsymbol{A} = (\{1, \ldots, n\}, \{a, b\}, \delta)$ where $\delta(i, a) = i+1$ if $i = 1, 2, \ldots, n-1$ and $\delta(n, a) = 1$, $\delta(n-1, b) = n$ and $\delta(i, b) = i$ if $i = 1, 2, \ldots, n$ and $i \neq n-1$.

Now we show that $d(\boldsymbol{A}) = (n-1)^2$. Let $w \in X^+$ be a shortest directing word of $\boldsymbol{A}$. Then $Sw^A = \bigcup_{s \in S}\{sw^A\} = \{n\}$ and $w = w'b$.

Moreover, $Sw'^A = \{n-1, n\}$. Let $w = w_1 a_1 a_2 \cdots a_{n-1} b$ where $a_i \in X, i = 1, 2, \ldots, n-1$. Then $a_i = a$ for any $i = 1, 2, \ldots, n-1$ and $Sw_1^A = \{n, 1\}$. This implies that $w_1 = w_1' b$ and $Sw_1'^A = \{n-1, n, 1\}$.

Taking the same procedure as above, we have:

$$w_1 = w_2 a^{n-1} b, Sw_2^A = \{n, 1, 2\} \text{ and } w_2 = w_2' b.$$

Continuing the same process, we can obtain $w = b(a^{n-1}b)^{n-2} = (ba^{n-1})^{n-2}b$. Since $|w| = (n-1)^2$, we have $(n-1)^2 \le d(n)$.

Proposition 8.1.3 *For any $n \ge 1$, we have $d(n) \le 1 + (n-2)\binom{n}{2}$.*

Proof It can easily be seen that the proposition holds for $n = 1, 2$. Let $n \ge 3$ and let $\boldsymbol{A} = (S, X, \delta)$ be a directable automaton with $|S| = n$. Since $\boldsymbol{A}$ is a directable automaton, there exist $s_0, s_1 \in S, s_0 \ne s_1$ and $a \in X$ such that $s_0 a^A = s_1 a^A$. Hence $|Sa^A| < |S|$. Let $s, t \in Sa^A, s \ne t$. Moreover, let $u \in X^*$ be one of the shortest words such that $su^A = tu^A$. Suppose $|u| \ge \binom{n}{2} + 1$. Then we have $u = xvy, v, xy \in X^+$ such that $sx^A = s(xv)^A$ and $tx^A = t(xv)^A$. This implies that $s(xy)^A = t(xy)^A$. This contradicts the assumption that u is one of the shortest words. Hence $|u| \le \binom{n}{2}$. Now consider $s', t' \in S(au)^A, s' \ne t'$. In the same way as the above, we can find $u' \in X^*, |u'| \le \binom{n}{2}$ such that $s'u'^A = t'u'^A$. Using the same technique, we can find a directing word $w = auu' \cdots$ of $\boldsymbol{A}$ with $|w| \le 1 + (n-2)\binom{n}{2}$.

A similar problem for some classes of automata can be disscussed. For instance, an automaton $\boldsymbol{A} = (S, X, \delta)$ is called a *commutative automaton* if $s(uv)^A = s(vu)^A$ holds for any $s \in S$ and any $u, v \in X^*$. By $d_{com}(n)$, we denote the value $max\{d(\boldsymbol{A}) \mid \boldsymbol{A} = (S, X, \delta)$ is commutative and directable, and $|S| = n\}$. In the definition of $d_{com}(n)$, X ranges over all finite nonempty alphabets. The following result is originally due to [60] and [61]. However, we have a simple proof due to [29].

Proposition 8.1.4 *For any $n \ge 1$, we have $d_{com}(n) = n - 1$.*

Proof Let $\boldsymbol{A} = (S, X, \delta)$ be a commutative directable automaton with $|S| = n$. First notice that $Su^A = \bigcup_{s \in S} su^A \supseteq S(vu)^A = S(uv)^A$ for any $u, v \in X^*$. Let $w = a_1 a_2 \cdots a_r$ be one of the shortest directing words of $\boldsymbol{A}$. Thus we have $Sa_1^A \supsetneq S(a_1a_2)^A \supsetneq S(a_1a_2a_3)^A \supsetneq \cdots \supsetneq S(a_1a_2a_3\cdots a_r)^A$. Notice that every inclusion is proper because the word w is one of the shotest directing words. Hence $r \leq n-1$. On the other hand, consider the following commutative directable automaton $\boldsymbol{B} = (\{1,2,\ldots,n\}, X, \gamma)$ where $\gamma(i,a) = i+1$ for any $a \in X$ and any $i = 1,2,\ldots,n-1$ and $\gamma(n,a) = n$ for any $a \in X$. Then $d(\boldsymbol{B}) = n-1$. Consequently, $d_{com}(n) = n-1$.

Let $\boldsymbol{A} = (S, X, \delta)$ be an automaton with $|S| = n$. We consider an algorithm to decide whether $\boldsymbol{A}$ is directable. Recall that a word $u \in X^*$ is called a *subword* of a word $w \in X^*$ if $w = xuy$ for some $x, y \in X^*$. Let $L \subseteq X^*$ and let $w \in X^*$. Then w is called a *merged word* of L if any word in L is a subword of w. Let $w \in X^*$ be a merged word of $X^{d(n)}$. Then an automaton $\boldsymbol{A} = (S, X, \delta)$ with $|S| = n$ is directable if and only if $|Sw^A| = 1$. Therefore, one of our purposes is to find a merged word of $X^{d(n)}$ which is as short as possible. For instance, $w = u_1u_2\cdots u_t$ is a merged word of $X^{d(n)}$ where $X^{d(n)} = \{u_1, u_2, \ldots, u_t\}$. However, this merged word is too long, i.e. $|w| = d(n)|X|^{d(n)}$. Actually, we have the following result [39]:

Proposition 8.1.5 *There exists a merged word $w \in X^+$ of $X^{d(n)}$ such that $|w| \leq |X|^{d(n)} + d(n) - 1$.*

To prove Proposition 8.1.5, we consider the automaton $\boldsymbol{C} = (X^{d(n)}, X, \gamma)$ where $(au)b^C = ub$ for $a, b \in X$ and $u \in X^{d(n)-1}$.

Let $v_1, v_2, \ldots, v_r \in X^{d(n)}$. Then, by $v_1 \to v_2 \to \cdots \to v_r$, we mean that $v_1, v_2, \ldots, v_r$ are distinct elements and $v_{i+1} = v_ia_i^C$ for some $a_i \in X$ where $i = 1,2,\ldots,r-1$.

Lemma 8.1.1 *Let $v_1 \to v_2 \to \cdots \to v_r$ for a positive integer r. If there is no $v \in X^{d(n)}$ such that $v_1 \to v_2 \to \cdots \to v_r \to v$, then $v_r \to v_1$.*

Proof Suppose that $v_r \to v_1$ does not hold. Let $v_i = a_iu_i$ where $a_i \in X$ and $u_i \in X^{d(n)-1}$ for any $i = 1,2,\ldots,r$. Since $v_r \to v_1$ does not hold, we have $\{v_ra^C \mid a \in X\} = \{u_ra \mid a \in X\} \subseteq \{v_2, v_3, \ldots, v_r\}$, i.e. $u_rX \subseteq \{v_2, v_3, \ldots, v_r\}$. Let $u_ra = v_i$ for some $a \in X$ and some

$i = 2, 3, \ldots, r$. By the fact that $v_{i-1} \to v_i$, $v_i \in u_{i-1}X$ and hence $u_{i-1} = u_r$. As $a_{i-1}u_{i-1} = v_{i-1}$, we have $v_{i-1} \in Xu_r$. That is, $v_i \in u_rX$ implies that $v_{i-1} \in Xu_r$. Notice that $|Xu_r| = |u_rX|$. Hence $Xu_r \subseteq \{v_1, v_2, \ldots v_{r-1}\}$. However, $v_r = a_ru_r \in Xu_r$. This is a contradiction. Therefore, $v_r \to v_1$ holds.

Lemma 8.1.2 *Let* $v_1 \to v_2 \to \cdots \to v_r$ *and let* $\{v_1, v_2, \ldots, v_r\} \neq X^{d(n)}$. *Then there exists* $v'_1 \to v'_2 \to \cdots \to v'_r \to v'_{r+1}$ *such that* $\{v_1, v_2, \ldots, v_r\} \subseteq \{v'_1, v'_2, \ldots, v'_r, v'_{r+1}\}$.

Proof Let $v \in X^{d(n)} \setminus \{v_1, v_2, \ldots, v_r\}$. Since $v \in X^{d(n)} \setminus \{v_1, v_2, \ldots, v_r\}$ and $vv_1^C = v_1$, there exist $u, w \in X^*$ and $a \in X$ such that $v_1 = uaw$ and $vu^C \in X^{d(n)} \setminus \{v_1, v_2, \ldots, v_r\}$ and $v(ua)^C = (vu^C)a^C \in \{v_1, v_2, \ldots, v_r\}$. Let $v_k = (vu^C)a^C$ with $k = 1, 2, \ldots, r$. Then $vu^C \to v_k$. On the other hand, it follows from Lemma 8.1.1 that $v_k \to v_{k+1} \to \cdots \to v_r \to v_1 \to v_2 \to \cdots \to v_{k-1}$. Recall that $vu^C \notin \{v_1, v_2, \ldots, v_r\}$. Thus $vu^C \to v_k \to v_{k+1} \to \cdots \to v_r \to v_1 \to v_2 \to \cdots \to v_{k-1}$. This completes the proof of the lemma.

Proof of Proposition 8.1.5 By Lemma 8.1.1 and Lemma 8.1.2, we have $v_1 \to v_2 \to \cdots \to v_t$ where $X^{d(n)} = \{v_1, v_2, \ldots, v_t\}$. Let $v_{i+1} = v_ia_i^C, a_i \in X, i = 1, 2, \ldots, t-1$. Consider the word $w = v_1a_1a_2 \cdots a_{t-1} \in X^*$. Obviously, v_1 is a subword of w. Since $v_1a_1^C = v_2$, v_2 is a subword of w as well. For any $i = 2, 3, \ldots, t$, v_i is a subword of w because $v_1(a_1a_2 \cdots a_{i-1})^C = v_i$. Hence w is a merged word of $X^{d(n)}$. Notice that $|w| = |v_1| + t - 1 = d(n) + |X|^{d(n)} - 1$. This completes the proof of the proposition.

Remark 8.1.1 It can easily be shown that the above word w is a shortest merged word of $X^{d(n)}$ since, for instance, v_1 can be found uniquely only at the prefix of w, v_2 can be found uniquely only at the suffix of v_1a_1 and in general v_i can be found uniquely only at the suffix of $v_1a_1a_2 \cdots a_{i-1}$.

Using the word w, we can check whether a given automaton $\boldsymbol{A} = (S, X, \delta)$ with $|S| = n$ is directable, i.e. $\boldsymbol{A}$ *is directable if and only if* $|Sw^C| = 1$.

However, the above algorithm is not effective. The following result [29] provides a more effective algorithm to decide whether a given automaton is directable.

Let ρ be the following relation on S:

$$\forall s, t \in S, (s,t) \in \rho \Longleftrightarrow \exists u \in X^*, su^A = tu^A.$$

We introduce also the following relations on S:

(1) $\forall s, t \in S, (s,t) \in \rho_0 \Longleftrightarrow s = t$.
(2) $\forall i \geq 1, \forall s, t \in S, (s,t) \in \rho_i \Longleftrightarrow \exists a \in X, (sa^A, ta^A) \in \rho_{i-1}$.

Proposition 8.1.6 $\rho = \bigcup_{i=0}^{\binom{n}{2}} \rho_i$. *Moreover,* $\boldsymbol{A}$ *is directable if and only if* $(s,t) \in \rho$ *for any* $s, t \in S$.

Proof That $\bigcup_{i=0}^{\binom{n}{2}} \rho_i \subseteq \rho$ is obvious. Let $(s,t) \in \rho$. Then there exists $u \in X^*$ such that $su^A = tu^A$. Assume that u is a shortest word such. Notice that $|\{T \mid T \subseteq S, |T| = 2\}| = \binom{n}{2}$. Therefore, $u = u_1u_2u_3$ where $u_1, u_2, u_3 \in X^*, u_2 \in X^+$ and $su_1^A = s(u_1u_2)^A = tu_1^A = t(u_1u_2)^A$. Thus $s(u_1u_3)^A = t(u_1u_3)^A$. This contradicts the minimality of $|u|$. Therefore, $|u| \leq \binom{n}{2}$, i.e. $u \in \bigcup_{i=0}^{\binom{n}{2}} \rho_i$. The latter half is obvious from the fact that $(s,t) \in \rho$ means $su^A = tu^A$ for some $u \in X^*$.

Remark 8.1.2 In [29], it is indicated that the above proposition provides an algorithm to decide whether or not any given automaton $\boldsymbol{A} = (S, X, \delta)$ is directable with the time bound of $\mathcal{O}(m \cdot n^2)$ where $m = |X|$ and $n = |S|$.

8.2 Nondeterministic case

In this section, we will deal with nondeterministic directable automata and their related languages. For nondeterministic automata, the directability can be defined in several ways. In each case, the directing words constitute a regular language. We will consider six classes of regular languages with respect to the different definitions of directability.

Notice that an automaton $\boldsymbol{A} = (S, X, \delta)$ is said to be *nondeterministic* if it consists of the following data: (1) S, X are the same materials as in the definition of finite automaton. (2) δ is a relation such that $\delta(s, a) \subseteq S$ for any $s \in S$ and $a \in X$.

In the above, the relation δ can be extended as follows: (1) $\delta(s, \epsilon) = \{s\}$ for any $s \in S$. (2) $\delta(s, av) = \bigcup_{t \in \delta(s,a)} \delta(t, v)$ for any $s \in S, a \in X$ and $v \in X^*$.

Let $\boldsymbol{A} = (S, X, \delta)$ be a nondeterministic automaton. First we introduce the notion of directing words of $\boldsymbol{A}$ by [30].

Definition 8.2.1 (1) A word $w \in X^*$ is D_1-*directing* if $sw^A \neq \emptyset$ for any $s \in S$ and $|Sw^A| = 1$. (2) A word $w \in X^*$ is D_2-*directing* if $sw^A = Sw^A$ for any $s \in S$. (3) A word $w \in X^*$ is D_3-*directing* if $\bigcap_{s \in S} sw^A \neq \emptyset$.

Definition 8.2.2 Let $i = 1, 2, 3$. Then $\boldsymbol{A}$ is called a D_i-*directable automaton* if the set of D_i-directing words is not empty.

Let $\boldsymbol{A} = (S, X, \delta)$ be a nondeterministic automaton. Then, for any $i = 1, 2, 3$, $\mathrm{D}_i(\boldsymbol{A})$ denotes the set of all D_i-directing words. Then we have:

Proposition 8.2.1 *For any $i = 1, 2, 3, \mathrm{D}_i(\boldsymbol{A})$ is a regular language.*

Proof Let $\boldsymbol{A} = (S, X, \delta)$ be a nondeterministic automaton and let $S = \{s_0, s_1, s_2, \ldots, s_r\}, r \geq 0$. For any $i = 1, 2, 3$, we define a finite acceptor $\boldsymbol{A}_i = (\{T \mid T \subseteq S\}, \delta_r, (\{s_0\}, \{s_1\}, \ldots, \{s_r\}), F_i)$ as follows: (1) $(T_0, T_1, \ldots, T_r)a^{A_i} = \delta_r((T_0, T_1, \ldots, T_r), a) = (T_0 a^A, T_1 a^A, \ldots, T_r a^A)$ for any $a \in X$ and any $T_i \subseteq S, i = 0, 1, \ldots, r$. (2) $F_1 = \{(\{t\}, \{t\}, \ldots, \{t\}) \mid t \in S\}$, $F_2 = \{(T, T, \ldots, T) \mid T \subseteq S\}$ and $F_3 = \{(T_0, T_1, \ldots, T_r) \mid T_i \subseteq S, i = 0, 1, \ldots, r, \bigcap_{i=0}^{r} T_i \neq \emptyset\}$.

Then it is obvious that $\mathrm{D}_i(\boldsymbol{A}) = \mathcal{T}(\boldsymbol{A}_i)$ for any $i = 1, 2, 3$. Hence $\mathrm{D}_i(\boldsymbol{A})$ is a regular language for any $i = 1, 2, 3$.

As for the D_1-directability of a complete nondeterministic automaton, Burkhard introduced it in [10]. We are going to investigate the classes of languages consisting of D_1-, D_2- and D_3-directing words of nondeterministic automata and complete nondeterministic automata. Here a *complete nondeterminsitic automaton* $\boldsymbol{A} = (S, X, \delta)$ means that $sa^{\boldsymbol{A}} \neq \emptyset$ for any $s \in S$ and any $a \in X$.

The classes of D_i-directable nondeterministic automata and complete nondeterministic automata are denoted by $\mathbf{Dir}(i)$ and $\mathbf{CDir}(i)$, respectively. Let X be an alphabet. For $i = 1, 2, 3$, we define the following classes of languages:

(1) $\mathcal{L}^X_{\mathrm{ND}(i)} = \{\mathrm{D}_i(\boldsymbol{A}) \mid \boldsymbol{A} = (S, X, \delta) \in \mathbf{Dir}(i)\}$. (2) $\mathcal{L}^X_{\mathrm{CND}(i)} = \{\mathrm{D}_i(\boldsymbol{A}) \mid \boldsymbol{A} = (S, X, \delta) \in \mathbf{CDir}(i)\}$.

Let $\mathbf{D}$ be the class of deterministic directable automata. For $\boldsymbol{A} \in \mathbf{D}$, $\mathrm{D}(\boldsymbol{A})$ denotes the set of all directing words of $\boldsymbol{A}$. Then we can define the class, i.e. $\mathcal{L}^X_{\mathrm{D}} = \{\mathrm{D}(\boldsymbol{A}) \mid \boldsymbol{A} = (S, X, \delta) \in \mathbf{D}\}$.

Then, by Propsition 8.2.1 and Proposition 8.1.1, all the above classes are subclasses of all regular languages.

8.3 Classes of regular languages

In this section, we will investigate the classes of regular languages consisting of directing words. Since it is easy to see that $\mathcal{L}^{\{a\}}_{\mathrm{D}} = \mathcal{L}^{\{a\}}_{\mathrm{CND}(i)} = \mathcal{L}^{\{a\}}_{\mathrm{ND}(i)} = \{a^n a^* \mid n \geq 1\}$ for any $i = 1, 2, 3$, we assume that $|X| \geq 2$.

We introduce the following result due to [30].

Lemma 8.3.1 *Let* $\boldsymbol{A} = (S, X, \delta)$ *be a nondeterministic automaton. Then we have* $\mathrm{D}_2(\boldsymbol{A})X^* = \mathrm{D}_2(\boldsymbol{A})$. *Moreover, If* $\boldsymbol{A}$ *is complete, then* $X^*\mathrm{D}_1(\boldsymbol{A}) = \mathrm{D}_1(\boldsymbol{A})$, $X^*\mathrm{D}_2(\boldsymbol{A})X^* = \mathrm{D}_2(\boldsymbol{A})$ *and* $X^*\mathrm{D}_3(\boldsymbol{A})X^* = \mathrm{D}_3(\boldsymbol{A})$.

Proof Let $\boldsymbol{A} = (S, X, \delta)$ be a nondeterministic automaton. First we prove that $\mathrm{D}_2(\boldsymbol{A})X^* = \mathrm{D}_2(\boldsymbol{A})$. It is obvious that $\mathrm{D}_2(\boldsymbol{A}) \subseteq \mathrm{D}_2(\boldsymbol{A})X^*$. Now, let $u \in \mathrm{D}_2(\boldsymbol{A})X^*$. Then $u = vw$ where $v \in \mathrm{D}_2(\boldsymbol{A})$ and $w \in X^*$. Let $s \in S$. Since $sv^A = Sv^A$, we have $su^A = s(vw)^A = (sv^A)w^A = (Sv^A)w^A = S(vw)^A = Su^A$. This means that $u \in \mathrm{D}_2(\boldsymbol{A})$. Hence $\mathrm{D}_2(\boldsymbol{A})X^* \subseteq \mathrm{D}_2(\boldsymbol{A})$ and $\mathrm{D}_2(\boldsymbol{A})X^* = \mathrm{D}_2(\boldsymbol{A})$.

Now assume $\boldsymbol{A}$ is complete. It is obvious that $\mathrm{D}_1(\boldsymbol{A}) \subseteq X^*\mathrm{D}_1(\boldsymbol{A})$. Let $u \in X^*\mathrm{D}_1(\boldsymbol{A})$. Then $u = vw$ where $v \in X^*$ and $w \in \mathrm{D}_1(\boldsymbol{A})$. Since $\boldsymbol{A}$ is complete, $su^A \neq \emptyset$ and $su^A = s(vw)^A = (sv^A)w^A \subseteq Sw^A$ for any $s \in S$. Consequently, $1 \leq |su^A| \leq |Sw^A| = 1$. Therefore, $|su^A| = 1$ and $u \in \mathrm{D}_1(\boldsymbol{A})$. This shows that $X^*\mathrm{D}_1(\boldsymbol{A}) = \mathrm{D}_1(\boldsymbol{A})$.

We show that $X^*\mathrm{D}_2(\boldsymbol{A})X^* = \mathrm{D}_2(\boldsymbol{A})$. Notice that $\mathrm{D}_2(\boldsymbol{A})X^* = \mathrm{D}_2(\boldsymbol{A})$ holds and hence $X^*\mathrm{D}_2(\boldsymbol{A})X^* = X^*\mathrm{D}_2(\boldsymbol{A})$ holds. Therefore, it is enough to show that $X^*\mathrm{D}_2(\boldsymbol{A}) = \mathrm{D}_2(\boldsymbol{A})$. That $\mathrm{D}_2(\boldsymbol{A}) \subseteq X^*\mathrm{D}_2(\boldsymbol{A})$ is obvious. Let $u \in X^*\mathrm{D}_2(\boldsymbol{A})$. Then $u = vw$ where $v \in X^*$ and $w \in \mathrm{D}_2(\boldsymbol{A})$. Let $s \in S$. As $su^A = s(vw)^A = (sv^A)w^A$, $sv^A \neq \emptyset$ and $w \in \mathrm{D}_2(\boldsymbol{A})$, $su^{\boldsymbol{A}} = (sv^A)w^{\boldsymbol{A}} = Sw^{\boldsymbol{A}}$. Notice that s is an arbitrary element of S. Therefore, $su^A = Su^A$ for any $s \in S$, i.e. $u \in \mathrm{D}_2(\boldsymbol{A})$. Thus $X^*\mathrm{D}_2(\boldsymbol{A}) = \mathrm{D}_2(\boldsymbol{A})$ and $X^*\mathrm{D}_2(\boldsymbol{A})X^* = \mathrm{D}_2(\boldsymbol{A})$.

We show the last equality. It is enogh to show that $X^*\mathrm{D}_3(\boldsymbol{A})X^* \subseteq \mathrm{D}_3(\boldsymbol{A})$. Let $u \in X^*\mathrm{D}_3(\boldsymbol{A})X^*$. Then $u = vwx$ where $v, x \in X^*$ and $w \in \mathrm{D}_3(\boldsymbol{A})$. Let $s \in S$. Since $\boldsymbol{A}$ is complete, $sv^A \neq \emptyset$. Let $sv^A = T$. Then $s(vw)^A = (sv^A)w^A = Tw^A \supseteq \bigcap_{t \in T} tw^A \supseteq \bigcap_{t \in S} tw^A \neq \emptyset$ because $w \in \mathrm{D}_3(\boldsymbol{A})$. Hence $s(vwx)^A \supseteq (\bigcap_{t \in S} tw^A)x^A \neq \emptyset$. Therefore, $\bigcap_{s \in S} su^A \neq \emptyset$. This means that $u \in \mathrm{D}_3(\boldsymbol{A})$. Consequently, we have $X^*\mathrm{D}_3(\boldsymbol{A})X^* = \mathrm{D}_3(\boldsymbol{A})$.

Now we will characterize the class $\mathcal{L}_{\mathrm{D}}^X$.

Proposition 8.3.1 *Let $L \subseteq X^*$ be a language. Then $L \in \mathcal{L}_{\mathrm{D}}^X$ if and only if $L \neq \emptyset$, L is regular and $X^*LX^* = L$.*

Proof $(\Rightarrow)$ Let $L \in \mathcal{L}_{\mathrm{D}}^X$. Then $L \neq \emptyset$ and there exists a directable automaton $\boldsymbol{A} = (S, X, \delta)$ such that $L = \mathrm{D}(\boldsymbol{A})$. Let $u \in L$, i.e. $u \in \mathrm{D}(\boldsymbol{A})$. Then $|Su^A| = 1$. Let $v, w \in X^*$. Since $\emptyset \neq S(vuw)^{\boldsymbol{A}} \subseteq$

$S(uw)^{\boldsymbol{A}} = (Su^{\boldsymbol{A}})w^{\boldsymbol{A}}$, we have $1 \leq |S(vuw)^{\boldsymbol{A}}| \leq |Su^{\boldsymbol{A}}| = 1$. Hence $|S(vuw)^{\boldsymbol{A}}| = 1$. This means that $vuw \in \mathrm{D}(\boldsymbol{A}) = L$, i.e. $X^*LX^* \subseteq L$. As the converse inclusion is obvious, we have $X^*LX^* = L$.

($\Leftarrow$) Now assume that $L \neq \emptyset$, L is regular and $X^*LX^* = L$. By the proof of Propositon 4.1, the accepter $\boldsymbol{A} = (S, X, \delta, [\epsilon], \{[u] \mid u \in L\})$ where $S = \{[u] \mid u \in X^*\}$ accepts L. Let $u, v \in L$. Then $xuy, xvy \in L$ for any $x, y \in X^*$ because $X^*LX^* = L$. That is, $u \equiv v(P_L)$ and $\{[u] \mid u \in L\}$ is a singleton set. Now let $w \in L$. Then $Sw^A = \{[uw] \mid u \in X^*\} = [w]$ because $uw \in L$ for any $u \in X^*$. Thus $|Sw^A| = 1$ and w is a directiong word of $\boldsymbol{A}$. Hence $L \in \mathcal{L}_{\mathrm{D}}^X$.

Corollary 8.3.1 *$\mathcal{L}_{\mathrm{D}}^X$ is closed under union.*

Proposition 8.3.2 *$\mathcal{L}_{\mathrm{CND(2)}}^X = \mathcal{L}_{\mathrm{D}}^X$, $\mathcal{L}_{\mathrm{CND(3)}}^X = \mathcal{L}_{\mathrm{D}}^X$, $\mathcal{L}_{\mathrm{CND(1)}}^X \cap \mathcal{L}_{\mathrm{ND(2)}}^X = \mathcal{L}_{\mathrm{D}}^X$ and $\mathcal{L}_{\mathrm{CND(1)}}^X \cap \mathcal{L}_{\mathrm{ND(3)}}^X = \mathcal{L}_{\mathrm{D}}^X$.*

Proof It is obvious that $\mathcal{L}_{\mathrm{D}}^X \subseteq \mathcal{L}_{\mathrm{CND(2)}}^X$. Now let $L \in \mathcal{L}_{\mathrm{CND(2)}}^X$. Then $L \neq \emptyset$. Moreover, by Lemma 8.3.1, we have $X^*LX^* = L$. By Proposition 8.3.1, $L \in \mathcal{L}_{\mathrm{D}}^X$ and hence $\mathcal{L}_{\mathrm{CND(2)}}^X = \mathcal{L}_{\mathrm{D}}^X$.

It is also obvious that $\mathcal{L}_{\mathrm{D}}^X \subseteq \mathcal{L}_{\mathrm{CND(3)}}^X$. Let $L \in \mathcal{L}_{\mathrm{CND(3)}}^X$. Then $L \neq \emptyset$ and by Lemma 8.3.1, we have $X^*LX^* = L$. By Proposition 8.3.1, $L \in \mathcal{L}_{\mathrm{D}}^X$. Thus we have $\mathcal{L}_{\mathrm{CND(3)}}^X = \mathcal{L}_{\mathrm{D}}^X$.

To prove that $\mathcal{L}_{\mathrm{CND(1)}}^X \cap \mathcal{L}_{\mathrm{ND(2)}}^X = \mathcal{L}_{\mathrm{D}}^X$, it is enough to show that $\mathcal{L}_{\mathrm{CND(1)}}^X \cap \mathcal{L}_{\mathrm{ND(2)}}^X \subseteq \mathcal{L}_{\mathrm{D}}^X$. Let $L \in \mathcal{L}_{\mathrm{CND(1)}}^X \cap \mathcal{L}_{\mathrm{ND(2)}}^X$. Since $L \in \mathcal{L}_{\mathrm{ND(2)}}^X$, by Lemma 8.3.1 $LX^* = L$ and hence $X^*LX^* = X^*L$. By the assumption that $L \in \mathcal{L}_{\mathrm{CND(1)}}^X$, it follows from Lemma 8.3.1 that $X^*LX^* = X^*L = L$. Therefore, by Lemma 8.3.1 we have $L \in \mathcal{L}_{\mathrm{D}}^X$, i.e. $\mathcal{L}_{\mathrm{CND(1)}}^X \cap \mathcal{L}_{\mathrm{ND(2)}}^X = \mathcal{L}_{\mathrm{D}}^X$.

Finally, to prove that $\mathcal{L}_{\mathrm{CND(1)}}^X \cap \mathcal{L}_{\mathrm{ND(3)}}^X = \mathcal{L}_{\mathrm{D}}^X$, it is enough to show that $\mathcal{L}_{\mathrm{CND(1)}}^X \cap \mathcal{L}_{\mathrm{ND(3)}}^X \subseteq \mathcal{L}_{\mathrm{D}}^X$. Let $L \in \mathcal{L}_{\mathrm{CND(1)}}^X \cap \mathcal{L}_{\mathrm{ND(3)}}^X$. Since $L \in \mathcal{L}_{\mathrm{CND(1)}}^X$, by Lemma 8.3.1 we have $X^*L = L$. Moreover, there exists a nondeterministic automaton $\boldsymbol{A} = (S, X, \delta)$ such that $L = \mathrm{D}_3(\boldsymbol{A})$ because $L \in \mathcal{L}_{\mathrm{ND(3)}}^X$. We show that $\boldsymbol{A}$ is a complete nondeterministic automaton. Suppose $\boldsymbol{A}$ is not complete. Then there exist $a \in X$ and $s \in S$ such that $sa^A = \emptyset$. Let $u \in L$. Since $X^*L = L$, $au \in$

$L = \mathrm{D}_3(\boldsymbol{A})$. Therefore, au is D_3-directing word of $\boldsymbol{A}$. Consequently, $s(au)^{\boldsymbol{A}} \neq \emptyset$ and hence $sa^A \neq \emptyset$, which is a contradiction. Hence $\boldsymbol{A} \in \mathbf{CDir}(3)$. By Lemma 8.3.1, $X^*LX^* = L$ and $L \in \mathcal{L}_{\mathrm{D}}^X$, i.e. $\mathcal{L}_{\mathrm{CND}(1)}^X \cap \mathcal{L}_{\mathrm{ND}(3)}^X = \mathcal{L}_{\mathrm{D}}^X$.

Notice that we have the following inclusion relations.

Proposition 8.3.3 *(1)* $\mathcal{L}_{\mathrm{D}}^X \subset \mathcal{L}_{\mathrm{CND}(1)}^X \subset \mathcal{L}_{\mathrm{ND}(1)}^X$. *(2)* $\mathcal{L}_{\mathrm{D}}^X \subset \mathcal{L}_{\mathrm{ND}(1)}^X \cap \mathcal{L}_{\mathrm{ND}(3)}^X$.

Proof (1) It is obvious that $\mathcal{L}_{\mathrm{D}}^X \subseteq \mathcal{L}_{\mathrm{CND}(1)}^X \subseteq \mathcal{L}_{\mathrm{ND}(1)}^X$ holds. Let $a \in X$. Consider the following nondeterministic automaton $\boldsymbol{A} = (\{1,2\}, X, \delta)$ where $1a^A = \{1\}, 2a^A = \{1\}$ and, $1b^A = \{1,2\}$ and $2b^A = \{2\}$ for $b \in X \setminus \{a\}$. Then $\boldsymbol{A}$ is complete. Moreover, a is a D_1-directing word of $\boldsymbol{A}$. If $\mathrm{D}_1(\boldsymbol{A}) \in \mathcal{L}_{\mathrm{D}}^X$, then ab must be in $\mathrm{D}_1(\boldsymbol{A})$ for any $b \in X \setminus \{a\}$. However, $ab \notin \mathrm{D}_1(\boldsymbol{A})$, which is a contradiction. Thus $\mathcal{L}_{\mathrm{D}}^X \subset \mathcal{L}_{\mathrm{CND}(1)}^X$. Let $\boldsymbol{B} = (\{1,2\}, X, \gamma)$ where $1a^B = \{1\}, 2a^B = \{1\}$ and, $1b^B = \emptyset$ and $2b^B = \{2\}$ for any $b \in X \setminus \{a\}$. Then a is D_1-directing word of $\boldsymbol{B}$. By 8.3.1, ba must be in $\mathrm{D}_1(\boldsymbol{B})$ for any $b \in X \setminus \{a\}$ if $\mathrm{D}_1(\boldsymbol{B}) \in \mathcal{L}_{\mathrm{CND}(1)}^X$. However, $ba \notin \mathrm{D}_1(\boldsymbol{B})$, which is a contradiction. Hence, $\mathcal{L}_{\mathrm{CND}(1)}^X \subset \mathcal{L}_{\mathrm{ND}(1)}^X$.

(2) Let $a \in X$. Since $\mathcal{L}_{\mathrm{D}}^X \subseteq \mathcal{L}_{\mathrm{ND}(1)}^X$ and $\mathcal{L}_{\mathrm{D}}^X \subseteq \mathcal{L}_{\mathrm{ND}(3)}^X$, $\mathcal{L}_{\mathrm{D}}^X \subseteq \mathcal{L}_{\mathrm{ND}(1)}^X \cap \mathcal{L}_{\mathrm{ND}(3)}^X$. Let $\boldsymbol{C} = (\{1,2\}, X, \theta)$ where $1a^C = \{1,2\}, 2a^C = \{1,2\}$ and, $1b^C = \{1\}$ and $2b^C = \emptyset$ for any $b \in X \setminus \{a\}$. Then $\mathrm{D}_2(\boldsymbol{C}) = \mathrm{D}_3(\boldsymbol{C}) = aX^*$, i.e. $aX^* \in \mathcal{L}_{\mathrm{ND}(1)}^X \cap \mathcal{L}_{\mathrm{ND}(3)}^X$. But, if $aX^* \in \mathcal{L}_{\mathrm{D}}^X$, then, by Lemma 8.3.1 ba must be in aX^* for any $b \in X \setminus \{a\}$, which is a contradiction. Hence $\mathcal{L}_{\mathrm{D}}^X \subset \mathcal{L}_{\mathrm{ND}(1)}^X \cap \mathcal{L}_{\mathrm{ND}(3)}^X$.

We will show that the classes $\mathcal{L}_{\mathrm{ND}(1)}^X, \mathcal{L}_{\mathrm{ND}(2)}^X$ and $\mathcal{L}_{\mathrm{ND}(3)}^X$ are incomparable, i.e. there is no inclusion relation between any two of these three classes. To describe the situation, we consider the following three nondeterministic automata $\boldsymbol{A}, \boldsymbol{B}, \boldsymbol{C}$ with the same set of states $\{1,2\}$:

(1) $1a^A = \{1\} = 2a^A = \{1\}, 1b^A = \{1,2\} = 2b^A = \{1,2\}$ for any $b \in X \setminus \{a\}$. (2) $1a^B = \{2\}, 2a^B = \{1,2\}, 1b^B = \emptyset, 2b^B = \{1\}$ for any $b \in X \setminus \{a\}$. (3) $1a^C = \{2\}, 2a^C = \{1,2\}, 1b^C = \{1\}, 2b^C = \emptyset$ for any $b \in X \setminus \{a\}$.

Lemma 8.3.2 *(1)* $\emptyset \neq \mathrm{D}_1(\boldsymbol{A}) \notin \mathcal{L}^X_{\mathrm{ND}(2)}$ *and* $\mathrm{D}_1(\boldsymbol{A}) \notin \mathcal{L}^X_{\mathrm{ND}(3)}$. *(2)* $\emptyset \neq \mathrm{D}_2(\boldsymbol{B}) \notin \mathcal{L}^X_{\mathrm{ND}(1)}$ *and* $\mathrm{D}_2(\boldsymbol{B}) \notin \mathcal{L}^X_{\mathrm{ND}(3)}$. *(3)* $\emptyset \neq \mathrm{D}_3(\boldsymbol{C}) \notin \mathcal{L}^X_{\mathrm{ND}(1)}$ *and* $\mathrm{D}_3(\boldsymbol{C}) \notin \mathcal{L}^X_{\mathrm{ND}(2)}$.

Proof (1) It is obvious that $\mathrm{D}_1(\boldsymbol{A}) = X^*a$. By Lemma 8.3.1, $X^*a \notin \mathcal{L}^X_{\mathrm{ND}(2)}$. Now suppose $X^*a \in \mathcal{L}^X_{\mathrm{ND}(3)}$. Then there exists a nondeterministic automaton $\boldsymbol{D} = (S, X, \delta)$ such that $\mathrm{D}_3(\boldsymbol{D}) = X^*a$. Since $a, ba \in X^*a$ where $b \in X \setminus \{a\}$, we have $a, ba \in \mathrm{D}_3(\boldsymbol{D})$. Therefore, $sa^{\boldsymbol{D}}, sb^{\boldsymbol{D}} \neq \emptyset$ for any $s \in S$. Hence $bab \in \mathrm{D}_3(\boldsymbol{D}) = X^*a$, which is a contradiction. Consequently, $X^*a \notin \mathcal{L}^X_{\mathrm{ND}(3)}$.

(2) Let $b \in X \setminus \{a\}$. Then $aa, bb \in \mathrm{D}_2(\boldsymbol{B})$. Suppose $\mathrm{D}_2(\boldsymbol{B}) \in \mathcal{L}_{\mathrm{ND}(1)}$. Then there exists a nondeterministic automaton $\boldsymbol{D} = (S, X, \delta)$ such $\mathrm{D}_2(\boldsymbol{B}) = \mathrm{D}_1(\boldsymbol{D})$. Since $aa, bb \in \mathrm{D}_1(\boldsymbol{D})$, we have $sb^{\boldsymbol{D}} \neq \emptyset$ for any $s \in S$ and $baa \in \mathrm{D}_1(\boldsymbol{D})$. Thus $baa \in \mathrm{D}_2(\boldsymbol{B})$, which is a contradiction. Now suppose $\mathrm{D}_2(\boldsymbol{B}) \in \mathcal{L}^X_{\mathrm{ND}(3)}$. Then there exists a nondeterministic automaton $\boldsymbol{E} = (T, X, \gamma)$ such $\mathrm{D}_2(\boldsymbol{B}) = \mathrm{D}_3(\boldsymbol{E})$. Notice that $aa, bb \in \mathrm{D}_2(\boldsymbol{B}) = \mathrm{D}_3(\boldsymbol{E})$, we have $tb^{\boldsymbol{E}} \neq \emptyset$ for any $t \in T$ and $baa \in \mathrm{D}_3(\boldsymbol{E})$. Hence $baa \in \mathrm{D}_2(\boldsymbol{B})$, a contradiction. Consequently, $\mathrm{D}_2(\boldsymbol{B}) \notin \mathcal{L}^X_{\mathrm{ND}(3)}$.

(3) Let $b \in X \setminus \{a\}$. Then $a, aab \in \mathrm{D}_3(\boldsymbol{C})$. Now suppose $\mathrm{D}_3(\boldsymbol{C}) \in \mathcal{L}_{\mathrm{ND}(1)}$. Then there exists a nondeterministic automaton $\boldsymbol{D} = (S, X, \delta)$ such $\mathrm{D}_3(\boldsymbol{C}) = \mathrm{D}_1(\boldsymbol{D})$. Since $a \in \mathrm{D}_1(\boldsymbol{D})$, we have $|sa^{\boldsymbol{D}}| = 1$ and $s(ab)^{\boldsymbol{D}} = (sa^{\boldsymbol{D}})b^{\boldsymbol{D}} = ((sa^{\boldsymbol{D}})a^{\boldsymbol{D}})b^{\boldsymbol{D}} = s(aab)^{\boldsymbol{D}}$ for any $s \in S$. Thus $ab \in \mathrm{D}_1(\boldsymbol{D})$ and $ab \in \mathrm{D}_3(\boldsymbol{C})$, which is a contradiction. Hence $\mathrm{D}_3(\boldsymbol{C}) \notin \mathcal{L}_{\mathrm{ND}(1)}$. Now suppose $\mathrm{D}_3(\boldsymbol{C}) \in \mathcal{L}^X_{\mathrm{ND}(2)}$. Then there exists a nondeterministic automaton $\boldsymbol{E}$ such $\mathrm{D}_2(\boldsymbol{E}) = \mathrm{D}_3(\boldsymbol{C})$. Since $a \in \mathrm{D}_2(\boldsymbol{E})$, by Lemma 8.3.1 we have $ab \in \mathrm{D}_2(\boldsymbol{E})$. Thus $ab \in \mathrm{D}_3(\boldsymbol{C})$, which is a contradiction. Thus $\mathrm{D}_3(\boldsymbol{C}) \notin \mathcal{L}^X_{\mathrm{ND}(2)}$.

Notice that $\mathrm{D}_1(\boldsymbol{A}) \in \mathcal{L}^X_{\mathrm{CND}(1)}$ in the proof of (1) of the above lemma. Hence the following proposition and its corollary are immediate consequences of Lemma 8.3.2.

Proposition 8.3.4 *The classes* $\mathcal{L}^X_{\mathrm{ND}(1)}, \mathcal{L}^X_{\mathrm{ND}(2)}$ *and* $\mathcal{L}^X_{\mathrm{ND}(3)}$ *are incomparable.*

Corollary 8.3.2 *The classes* $\mathcal{L}^X_{\mathrm{CND}(1)}, \mathcal{L}^X_{\mathrm{ND}(2)}$ *and* $\mathcal{L}^X_{\mathrm{ND}(3)}$ *are incomparable.*

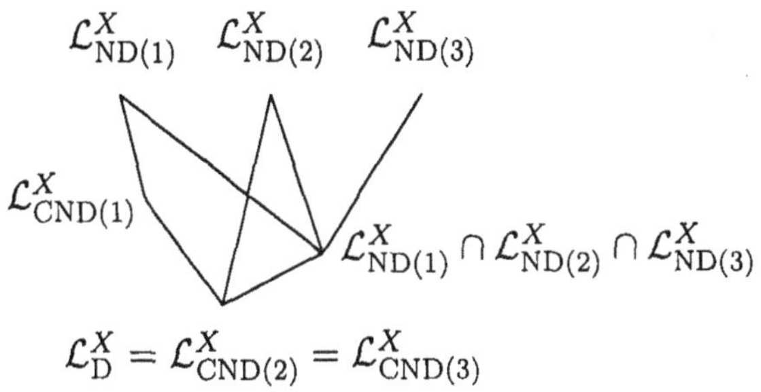

Figure 8.1: Inclusion relations

The above figure depicts the inclusion relations for the 7 classes of regular languages.

Now we investigate the structure of a language in $\mathcal{L}^X_{\mathrm{ND}(1)}$.

Definition 8.3.1 A nondeterministic automaton $\boldsymbol{A} = (S, X, \delta)$ is said to be *of partial function type* (or in short, *of pf-type*) if $|sa^A| \leq 1$ for any $s \in S$ and any $a \in X$.

Lemma 8.3.3 *Let $\boldsymbol{A} = (S, X, \delta)$ be a nondeterministic automaton. Then there exists a nondeterministic automaton of pf-type $\boldsymbol{B} = (T, X, \gamma)$ such that $\mathrm{D}_2(\boldsymbol{A}) = \mathrm{D}_2(\boldsymbol{B})$.*

Proof We define $\boldsymbol{B} = (\{Z \mid \emptyset \neq Z \subseteq S\}, X, \gamma)$ as follows:

$Za^B = \bigcup_{z \in Z} za^A$ for any $s \in S$ and any $Z \subseteq S, Z \neq \emptyset$.

Let $u \in \mathrm{D}_2(\boldsymbol{A})$. Then $su^A = Su^A$ for any $s \in S$. Therefore, $Zu^B = Su^B$ for any $Z \subseteq S, Z \neq \emptyset$. Thus $u \in \mathrm{D}_2(\boldsymbol{A})$.

Now assume $u \notin \mathrm{D}_2(\boldsymbol{A})$. Then there exist $s, t \in S$ such that $su^A \neq tu^A$. Hence $\{s\}u^B \neq \{t\}u^B$. This means that $u \notin \mathrm{D}_2(\boldsymbol{B})$. Hence $\mathrm{D}_2(\boldsymbol{A}) = \mathrm{D}_2(\boldsymbol{B})$.

Proposition 8.3.5 *Let $L \in \mathcal{L}^X_{\mathrm{ND}(2)} \setminus (\mathcal{L}^X_{\mathrm{ND}(1)} \cap \mathcal{L}^X_{\mathrm{ND}(3)})$. Then there exist $L_1 \in \mathcal{L}^X_{\mathrm{D}}, L_2 \in \mathcal{L}^X_{\mathrm{ND}(1)} \cap \mathcal{L}^X_{\mathrm{ND}(3)}$ such that $L = L_1 \cup L_2$ with $L_1 \cap L_2 \neq \emptyset$.*

Proof By Lemma 8.3.3, there exists a nondeterministic automaton of pf-type $\boldsymbol{A}$ such that $L = \mathrm{D}_2(\boldsymbol{A})$. Let $L_1 = \{u \in L \mid Su^A = \emptyset\}$ and let $L_2 = \{u \in L \mid Su^A \neq \emptyset\}$. Then it is obvious that $L_2 = \mathrm{D}_1(\boldsymbol{A}) = \mathrm{D}_3(\boldsymbol{A})$. Therefore, $L_2 \in \mathcal{L}^X_{\mathrm{ND}(1)} \cap \mathcal{L}^X_{\mathrm{ND}(3)}$ if $L_2 \neq \emptyset$. Notice that $L_1 = L \setminus L_2$ and L, L_2 are regular. Hence L_1 is regular. It is obvious that $X^*L_1X^* = L_1$. By Propositon 8.3.1, $L_1 \in \mathcal{L}^X_{\mathrm{D}}$ if $L_1 \neq \emptyset$. Suppose that $L_1 = \emptyset$ or $L_2 = \emptyset$. Then $L \in \mathcal{L}^X_{\mathrm{ND}(1)} \cap \mathcal{L}^X_{\mathrm{ND}(3)}$, which is a contradiction. This completes the proof of the proposition.

We have already proven that $\mathcal{L}^X_{\mathrm{D}}$ is closed under union. We will investigate some closure properties under union and intersection.

Proposition 8.3.6 *The following classes are not closed under union:*

(1) $\mathcal{L}^X_{\mathrm{CND}(1)}$. *(2)* $\mathcal{L}^X_{\mathrm{ND}(1)}$. *(3)* $\mathcal{L}^X_{\mathrm{ND}(2)}$.

Proof (1) Let $a \in X$ and let $A = (\{1,2,3\}, X, \delta)$ be the following complete nondeterministic automaton:

$1a^A = \{1\}, 2a^A = \{1\}, 3a^A = \{2\}, 1b^A = 2b^A = 2b^A = \{1,2,3\}$ for any $b \in X \setminus \{a\}$.

Then $\mathrm{D}_1(\boldsymbol{A}) = X^*a^2$, i.e. $X^*a^2 \in \mathcal{L}^X_{\mathrm{CND}(1)}$ for any $a \in X$. Consider $L = X^*a^2 \cup X^*b^2$ for $a, b \in X, a \neq b$.

Suppose $L \in \mathcal{L}^X_{\mathrm{ND}(1)}$. Then there exists a nondeterministic automaton $B = (S, X, \gamma)$ such that $L = \mathrm{D}_1(\boldsymbol{B})$. Notice that $a^2 \in \mathrm{D}_1(\boldsymbol{B})$. Since $a^2 \in \mathrm{D}_1(\boldsymbol{B}), sa^B = \emptyset$ for any $s \in S$. On the other hand, since $b^2 \in \mathrm{D}_1(\boldsymbol{B}), s(abb)^B = S(abb)^B$ for any $s \in S$. Thus $abb \in \mathrm{D}_1(\boldsymbol{B}) = L$, which is a contradiction. Consequently, $L \notin \mathcal{L}^X_{\mathrm{ND}(1)}$ and hence $L \notin \mathcal{L}^X_{\mathrm{CND}(1)}$, i.e. $\mathcal{L}^X_{\mathrm{CND}(1)}$ is not closed under union.

(2) By (1), $L \notin \mathcal{L}^X_{\mathrm{ND}(1)}$ and hence $\mathcal{L}^X_{\mathrm{ND}(1)}$ is not closed under union.

(3) Let $a \in X$ and let $C = (\{1,2,3\}, X, \delta)$ be the following nondeterministic automaton:

$1a^A = \{1\}, 2a^A = \{1\}, 3a^A = \{2\}, 1b^A = \{1\}, 2b^A = 3b^A = \emptyset$ for any $b \in X \setminus \{a\}$.

Then $\mathrm{D}_2(\boldsymbol{C}) = a^2X^*$, i.e. $a^2X^* \in \mathcal{L}^X_{\mathrm{ND}(2)}$. Consider $K = a^2X^* \cup b^2X^*$ for $a, b \in X, a \neq b$. Suppose $K \in \mathcal{L}^X_{\mathrm{ND}(2)}$. Then there exists a nondeterministic automaton $\boldsymbol{E} = (Q, X, \theta)$ such that $K = \mathrm{D}_2(\boldsymbol{E})$. If $Q(aa)^E = \emptyset$, then $Q(baa)^E = \emptyset$ and $baa \in \mathrm{D}_2(\boldsymbol{E}) = K$, which is a contradiction. If $Q(aa)^E \neq \emptyset$, then $q(aa)^E \neq \emptyset$ for any $q \in Q$ and hence $qa^E \neq \emptyset$ for any $q \in Q$. In this case, $q(abb)^E = Q(abb)^E$ for any $q \in Q$, i.e. $abb \in \mathrm{D}_2(\boldsymbol{E}) = K$, which is a contradiction. Therefore, $K \in \mathcal{L}^X_{\mathrm{ND}(2)}$, i.e. $\mathcal{L}^X_{\mathrm{ND}(2)}$ is not closed under union.

Proposition 8.3.7 *The class $\mathcal{L}^X_{\mathrm{ND}(3)}$ is closed under union.*

Proof Let $L, M \in \mathcal{L}^X_{\mathrm{ND}(3)}$. Then there exist nondeterministic automata $\boldsymbol{A} = (S, X, \delta)$ and $\boldsymbol{B} = (T, X, \gamma)$ such that $S \cap T = \emptyset, \mathrm{D}_3(\boldsymbol{A}) = L$ and $\mathrm{D}_3(\boldsymbol{B}) = K$. Now consider the following nondeterministic automaton $\boldsymbol{C} = (S \cup T, X, \theta)$:

$sa^C = sa^A \cup T$ for any $s \in S$ and any $a \in X$ and $ta^C = ta^B \cup S$ for any $t \in T$ and any $a \in X$.

We will show that $L \cup M = \mathrm{D}_3(\boldsymbol{C})$.

Let $u \in L$. Then $\bigcap_{s \in S} su^A \neq \emptyset$. Hence $\bigcap_{q \in S \cup T} qu^C = ((\bigcap_{s \in S} su^A) \cup T) \cap ((\bigcap_{t \in T} tu^B) \cup S) \supseteq \bigcap_{s \in S} su^A \neq \emptyset$. Thus $u \in \mathrm{D}_3(\boldsymbol{C})$. In the same way, if $u \in M$, then $u \in \mathrm{D}_3(\boldsymbol{C})$. Therefore, $L \cup M \subseteq \mathrm{D}_3(\boldsymbol{B})$.

Let $u \notin L \cup M$. Then $u \notin L$ and $u \notin M$. This implies that $\bigcap_{s \in S} su^A = \emptyset$ and $\bigcap_{t \in T} tu^B = \emptyset$. Hence $\bigcap_{q \in S \cup T} qu^C = ((\bigcap_{s \in S} su^A) \cup T) \cap ((\bigcap_{t \in T} tu^B) \cup S) = \bigcap_{s \in S} su^A \neq \emptyset$. Therefore, $u \notin \mathrm{D}_3(\boldsymbol{C})$ and $L \cup M = \mathrm{D}_3(\boldsymbol{C})$, i.e. $L \cup M \in \mathcal{L}^X_{\mathrm{ND}(3)}$. This completes the proof of the proposition.

Proposition 8.3.8 *The following classes are closed under nonempty intersection: (1) $\mathcal{L}^X_{\mathrm{D}}$. (2) $\mathcal{L}^X_{\mathrm{CND}(1)}$. (3) $\mathcal{L}^X_{\mathrm{ND}(1)}$. (4) $\mathcal{L}^X_{\mathrm{ND}(2)}$. (5) $\mathcal{L}^X_{\mathrm{ND}(3)}$.*

Outline of the proof Let $i = 1, 2, 3$ and let $\boldsymbol{A} = (S, X, \delta), \boldsymbol{B} = (T, X, \gamma)$ be nondeterministic automata such that $\mathrm{D}_i(\boldsymbol{A}) = L$ and $\mathrm{D}_i(\boldsymbol{B}) = K$. Define the nondeterministic automaton $\boldsymbol{A} \times \boldsymbol{B} = (S \times T, X, \delta \times \gamma)$ as follows:

$(s,t)a^{A\times B} = \{(s',t') \mid s' \in sa^A, t' \in ta^B\}$.

Notice that $\boldsymbol{A} \times \boldsymbol{B}$ is complete if and only if $\boldsymbol{A}$ and $\boldsymbol{B}$ are complete. For any $i = 1,2,3$, we can see that $u \in \mathrm{D}_i(\boldsymbol{A} \times \boldsymbol{B})$ if and only if $u \in \mathrm{D}_i(\boldsymbol{A}) \cap \mathrm{D}_i(\boldsymbol{B})$ for any $u \in X^*$. Thus any of the above classes is closed under nonempty intersection.

We will consider the shortest directing words of nondeterministic automata.

Let $i = 1,2,3$ and let $\boldsymbol{A} = (S,X,\delta)$ be a nondeterministic automaton. Then $d_i(\boldsymbol{A})$ denotes the value $min\{|u| \mid u \in \mathrm{D}_i(\boldsymbol{A})\}$. For any positive integer $n \geq 1$, $d_i(n)$ denotes the value $max\{d_i(\boldsymbol{A}) \mid \boldsymbol{A} = (S,X,\delta) : \boldsymbol{A} \in \mathbf{Dir}(i)$ and $|S| = n\}$. Moreover, $cd_i(n)$ denotes the value $max\{d_i(\boldsymbol{A}) \mid \boldsymbol{A} = (S,X,\delta) : \boldsymbol{A} \in \mathbf{CDir}(i)$ and $|S| = n\}$. Notice that in the definitions of $d_i(n)$ and $cd_i(n)$, X ranges over all finite nonempty alphabets.

First, we determine the value $cd_1(n)$. The following result is due to [10].

Theorem 8.3.1 *Let $n \geq 1$. Then $cd_1(n) = 2^n - n - 1$.*

Proof For $n = 1,2$, obviously the assertion of the theorem holds true. Assume $n \geq 3$.

Proof of $cd_1(n) \leq 2^n - n - 1$ Let $\boldsymbol{A} = (S,X,\delta)$ be a complete nondeterministic automaton with $|S| = n$ and let $w = a_1a_2\cdots a_{r-1}a$ $\in \mathrm{D}_1(\boldsymbol{A})$ where $a_1, a_2, \ldots, a_{r-1}, a \in X$ and $r = |w| = d_1(\boldsymbol{A})$. Suppose that $r = |w| > 2^n - n - 1$. Notice that $2^n - n - 2 = |\{T \subset S \mid |T| \geq 2\}|$. Let $T_i = S(a_1a_2\cdots a_i)^A$ where $i = 1,2,\ldots,r-1$, Then there exist $i,j = 1,2,\ldots,r-1$ such that $i < j$ and $T_i = T_j$. In this case, $a_1a_2\cdots a_ia_{j+1}\cdots a_ra \in \mathrm{D}_1(\boldsymbol{A})$. This contradicts the assumption that $r = d_1(\boldsymbol{A})$. Hence $cd_1(n) \leq 2^n - n - 1$.

Proof of $2^n - n - 1 \leq cd_1(n)$ To prove $2^n - n - 1 \leq cd_1(n)$, it is enough to construct a complete nondeterministic automaton $\boldsymbol{A} = (S,X,\delta)$ such that $|S| = n$ and $d_1(\boldsymbol{A}) = 2^n - n - 1$. Let S be a finite set with $|S| = n$. Moreover, let $\{T_1, T_2, \ldots, T_r\} = \{T \subset S \mid |T| \geq 2\}$. Furtheremore, we assume that $|T_1| \geq |T_2| \geq \cdots \geq |T_r|$ and

$T_r = \{s_1, s_2\}$. Notice that $r = 2^n - n - 2$. Now we construct the following nondeterministic automaton $\boldsymbol{A} = (S, X, \delta)$:

(1) $X = \{a_1, a_2, \ldots, a_r, z\}$. (2) $sa_1^A = T_1$ for any $s \in S$. (3) For any $i = 1, 2, \ldots, r-1$, $sa_{i+1}^A = T_{i+1}$ if $s \in T_i$ and $sa_{i+1}^A = S$, otherwise. (4) $s_1 z^A = s_2 z^A = \{s_1\}$ and $sz^A = S$ if $s \neq s_1, s_2$.

Then $\boldsymbol{A}$ is complete and $S(a_1 a_2 \cdots a_r z)^A = Sa_1^A(a_2 \cdots a_r z)^A = T_1(a_2 \cdots a_r z)^A = T_1 a_2^A (a_3 \cdots a_r z)^A = T_2(a_3 \cdots a_r z)^A = \cdots = T_r z^A = \{s_1, s_2\} z^A = \{s_1\}$. Hence $a_1 a_2 \cdots a_r z \in \mathrm{D}_1(\boldsymbol{A})$ and $|a_1 a_2 \cdots a_r z| = r + 1 = 2^n - n - 1$. Suppose that there exists $w \in \mathrm{D}_1(\boldsymbol{A})$ with $d_1(\boldsymbol{A}) = |w| < 2^n - n - 1$. Let $w = a_i w'$ where $w' \in X^*$ and $i = 2, 3, \ldots, r$. Since $S \supset T_{i-1}$, we have $Sa_i = S$, $S(a_i w')^A = Sw'^A$ and $|Sw'^A| = 1$. This implies that $w' \in \mathrm{D}_1(\boldsymbol{A})$. This contradicts the minimality of $|w|$. For the case that $w = zw'$ where $w' \in X^*$, we have $S(zw')^A = Sz^A w'^A = Sw'^A$ because $S \supset \{s_1, s_2\}$. Hence we encounter the same contradiction. Assume $w = a_1 a_2 \cdots a_i w'$ for some $i = 1, 2, \ldots, r$ and $w' \in X^*$. Obviously, $w' \in X^+$ and hence $i \leq r - 1$. Suppose $w' = a_j v$ where $j \neq i + 1$ and $v \in X^*$. Let $j < i + 1$. Remark that $w = a_1 a_2 \cdots a_j \cdots a_i a_j v$. Notice that $S(a_1 a_2 \cdots a_j \cdots a_i a_j)^A = T_j$ or $S(a_1 a_2 \cdots a_j \cdots a_i a_j)^A = S$. In the former case, since $S(a_1 a_2 \cdots a_j)^A = T_j$, we have $S(a_1 a_2 \cdots a_j v)^A = Sw^A$ and $a_1 a_2 \cdots a_j v \in \mathrm{D}_1(\boldsymbol{A})$, which contradicts the minimality of $|w|$. In the latter case, $Sw^A = Sv^A$ and $|Sv^A| = 1$. Therefore, $v \in \mathrm{D}_1(\boldsymbol{A})$, which is a contradiction. Suppose $j > i + 1$. Since $i < j - 1$, $T_i \setminus T_{j-1} \neq \emptyset$. Therefore, $S(a_1 a_2 \cdots a_i a_j)^A = T_i a_j^A = S$. Consequently, $Sv^A = Sw^A$ and $v \in \mathrm{D}_1(\boldsymbol{A})$, which contradicts the minimality of $|w|$. Now let $w' = zv$ where $v \in X^*$. Since $w = a_1 a_2 \cdots a_i z v$, $Sw^A = S(a_1 a_2 \cdots a_i z v)^A = T_i z^A v^A$. Notice that $T_i \neq T_r$. Hence $T_i \setminus \{s_1, s_2\} \neq \emptyset$. Therefore, $Sw^A = S(a_1 a_2 \cdots a_i z v)^A = T_i z^A v^A = Sv^A$ and $v \in \mathrm{D}_1(\boldsymbol{A})$, which contradicts the minimality of $|w|$. This means that there is no $w \in \mathrm{D}_1(\boldsymbol{A})$ with $|w| < 2^n - n - 1$. Hence $d_1(\boldsymbol{A}) = 2^n - n - 1$. This completes the proof of the proposition.

We will first deal with the value $d_1(n)$ and then the value $d_3(n)$. The results are due to [48].

Lemma 8.3.4 *Let* $n \geq 2$. *Then* $d_1(n) \leq \sum_{k=2}^{n} \binom{n}{k}(2^k - 1)$.

Proof Let $n \geq 2$. We will show that $d_1(n) \leq \sum_{k=2}^{n} \binom{n}{k}(2^k - 1)$. Let $\boldsymbol{A} = (S, X, \delta)$ be a D_1-directable automaton with n states and let $w = a_1 a_2 \cdots a_r \in \mathrm{D}_1(\boldsymbol{A})$ such that $a_i \in X, i = 1, 2, \ldots, r, r \geq 1$ and $|w| = r = d_1(\boldsymbol{A})$. Since $w \in \mathrm{D}_1(\boldsymbol{A})$, there exists $s_0 \in S$ such that $sw^{\boldsymbol{A}} = \{s_0\}$ for any $s \in S$. For any $i = 1, 2, \ldots, r$, we define the set S_i and T_i as follows: (1) $S_i = S(a_1 a_2 \cdots a_i)^{\boldsymbol{A}}$. (2) $T_i = \{t \in S_i \mid t(a_{i+1} a_{i+2} \cdots a_r)^{\boldsymbol{A}} = \{s_0\}\}$.

Let $s \in S$ and let $i = 1, 2, \ldots, r$. Since $s(a_1 a_2 \cdots a_i a_{i+1} \cdots a_r)^{\boldsymbol{A}} = (s(a_1 a_2 \cdots a_i)^{\boldsymbol{A}})(a_{i+1} \cdots a_r)^{\boldsymbol{A}} = \{s_0\}$, we have $s(a_1 a_2 \cdots a_i)^{\boldsymbol{A}} \cap T_i \neq \emptyset$. Let $S = S_0 = T_0$. Consider the set $\{(S_i, T_i) \mid i = 0, 1, 2, \ldots, r-1\}$. It is obvious that $S_i \neq \emptyset$ for any $i = 0, 1, \ldots, r-1$. It is also obvious that $|S_0| \neq 1$. Suppose that $|S_i| = 1$ for some $i = 1, 2, \ldots, r-1$. Then $S_i = T_i = \{t\}$ for some $t \in S$. By the definition of T_i, this means that $a_{i+1} a_{i+2} \cdots a_r \in \mathrm{D}_1(\boldsymbol{A})$, which contradicts the minimality of $|w|$. Therefore, $|S_i| \neq 1$ for any $i = 1, 2, \ldots, n$. Hence the set $\{(S_i, T_i) \mid i = 0, 1, 2, \ldots, r-1\}$ does not contain any $(\{s\}, \{s\})$ with $s_0 \neq s \in S$.

Now assume that $(S_i, T_i) = (S_j, T_j)$ for some $i, j = 1, 2, \ldots, r-1, i < j$. Then it can be seen that $a_1 a_2 \cdots a_i a_{j+1} a_{j+2} \cdots a_r \in \mathrm{D}_1(\boldsymbol{A})$, which contradicts the minimality of $|w|$. Hence all $(S_i, T_i), i = 0, 1, 2, \ldots, r-1$, are distinct. Therefore, $|\{(S_i, T_i) \mid i = 0, 1, 2, \ldots, r-1\}| \leq \sum_{k=2}^{n} \binom{n}{k}(2^k - 1)$ and hence $r \leq \sum_{k=2}^{n} \binom{n}{k}(2^k - 1)$. This completes the proof of the lemma.

To determine the value $d_1(2)$, we consider the following example.

Example 8.3.1 Let $\boldsymbol{A} = (\{1, 2\}, \{a, b, c\}, \delta)$ be the following nondeterministic automaton: (1) $1a^{\boldsymbol{A}} = \{1, 2\}$ and $2a^{\boldsymbol{A}} = \{2\}$. (2) $1b^{\boldsymbol{A}} = \emptyset$ and $2b^{\boldsymbol{A}} = \{1, 2\}$. (3) $1c^{\boldsymbol{A}} = \{1\}$ and $2c^{\boldsymbol{A}} = \emptyset$.

Then abc is a shortest D_1-directing word of $\boldsymbol{A}$. By Lemma 8.3.4, $d_1(2) \leq 3$. Therefore, $d_1(2) = 3$.

Lemma 8.3.5 *Let $n \geq 3$. Then $2^n - n \leq d_1(n)$.*

Proof For $n = 1, 2$, obviously the lemma holds true. Assume $n \geq 3$. We will construct a D_1-directable automaton $\boldsymbol{A} = (S, X, \delta)$ such that

$|S| = n$ and $d_1(\boldsymbol{A}) = 2^n - n$. Let S be a finite set with $|S| = n$ and let $\{T_1, T_2, \ldots, T_r\} = \{T \subset S \mid |T| \geq 2\}$. Notice that $r = 2^n - n - 2$. Moreover, we assume that $|T_1| \geq |T_2| \geq \cdots \geq |T_r|$, $\{s_0\} = S \setminus T_1$ and $T_r = \{s_1, s_2\}$. Now we construct the following nondeterministic automaton $\boldsymbol{A} = (S, X, \delta)$: (1) $X = \{a_1, a_2, \ldots, a_r, b\}$. (2) For any $i = 1, 2, \ldots, r-1$, $sa_i^A = T_{i+1}$ if $s \in T_i$ and $sa_i^A = S$, otherwise. (3) $s_1 a_r^A = s_2 a_r^A = \{s_1\}$ and $sa_r^A = S$ if $s \in S \setminus \{s_1, s_2\}$. (4) $s_0 b^A = \emptyset$ and $sb^A = T_1$ for any $s \in S \setminus \{s_0\}$.

Let $s \in S$ and let $i = 1, 2, \ldots, r$. Notice that $s(a_i b a_1 a_2 \cdots a_r)^A = \{s_1\}$ and hence $a_i b a_1 a_2 \cdots a_r \in \mathrm{D}_1(\boldsymbol{A})$. Moreover, since $s_0 b^A = \emptyset$, we have $bX^* \cap \mathrm{D}_1(\boldsymbol{A}) = \emptyset$. Let $i, j = 1, 2, \ldots, r$. Then $S(a_i a_j)^A = S$. On the other hand, $s(a_i b)^A = T_1$ for any $s \in S$. This means that $u \in a_i b X^*$ if u is a shortest D_1-directing word of $\boldsymbol{A}$. Let $i = 1, 2, \ldots, r-1$. Then $T_i(a_i a_j)^A = T_{i+1} a_j^A = S$ if $j > i+1$ and $T_i(a_i a_j)^A = T_{i+1} a_j^A = T_{j+1}$ if $j \leq i$. Notice that in the latter case $j+1 \leq i+1$. This implies that u is not a shortest D_1-directing word of $\boldsymbol{A}$ if $u \in X^* a_i a_j X^*$ where $j \neq i+1$. Moreover, since $Sb^A = T_1$, u is not a shortest D_1-directing word of $\boldsymbol{A}$ if $u \in XX^+ bX^*$. Consequently, $a_i b a_1 a_2 \cdots a_r$ is a shortest D_1-directing word of $\boldsymbol{A}$, i.e. $d_1(\boldsymbol{A}) = r + 2 = 2^n - n$. This completes the proof of the lemma.

Thus we have the following.

Proposition 8.3.9 *Let $n \geq 2$. Then $2^n - n \leq d_1(n) \leq \sum_{k=2}^{n} \binom{n}{k}(2^k - 1)$. Notice that $d_1(1) = 0$ and $d_1(2) = 3$.*

Before dealing with the value $d_3(n)$, recall that a nondeterministic automaton $\boldsymbol{A} = (S, X, \delta)$ is said to be *of pf-type* if $|sa^A| \leq 1$ for any $s \in S$ and any $a \in X$.

Remark 8.3.1 Let $\boldsymbol{A}$ be a nondeterministic automaton of *pf*-type. Then $\mathrm{D}_3(\boldsymbol{A}) = \mathrm{D}_1(\boldsymbol{A})$.

Let $\boldsymbol{A} = (S, X, \delta)$ be a D_3-directable automaton of *pf*-type. Consider the following procedure $\mathcal{P}$: Let $u \in \mathrm{D}_3(\boldsymbol{A})$. Assume that $u = u_1 u_2 u_3$ where $u_1, u_3 \in X^*, u_2 \in X^+$ and $Su_1{}^A = S(u_1 u_2)^A$. Then procedure $\mathcal{P}$ can be applied as $u \Rightarrow^{\mathcal{P}} u_1 u_3$.

Then we have the following result.

Lemma 8.3.6 *In the above procedure, we have* $u_1u_3 \in \mathrm{D}_3(\boldsymbol{A})$.

Proof Let $\boldsymbol{A} = (S, X, \delta)$ be a nondeterministic automaton of *pf*-type. Moreover, let $u = u_1u_2u_3$ where $u_1, u_3 \in X^*, u_2 \in X^+$ and $Su_1{}^A = S(u_1u_2)^A$. Since $u \in \mathrm{D}_3(\boldsymbol{A})$, there exists $s_0 \in S$ such that $su^A = \{s_0\}$ for any $s \in S$. From the assumptions that $Su_1{}^A = S(u_1u_2)^A$ and $\boldsymbol{A}$ is a nondeterministic automaton of *pf*-type, it follows that $su^A = s(u_1u_2u_3)^A = s(u_1u_3)^A = \{s_0\}$ for any $s \in S$. By Remark 8.3.1, this means that $u_1u_3 \in \mathrm{D}_3(\boldsymbol{A})$.

Let $\boldsymbol{A} = (S, X, \delta)$ be a D_3-directable automaton of *pf*-type and let $a_1a_2\cdots a_r \in \mathrm{D}_3(\boldsymbol{A})$ such that $\exists s, t \in S, s \neq t, sa_1{}^A = ta_1{}^A$. Assume that $v \in \mathrm{D}_3(\boldsymbol{A}), v = v_1v_2v_3, v_1, v_3 \in X^*, v_2 \in X^+$, $|Sv_1{}^A| = |S(v_1v_2)^A|$ and $\{s, t\} \subseteq Sv_1{}^A$. Then procedure $\mathcal{Q}_{(s,t)}$ can be applied as $v \Rightarrow^{\mathcal{Q}_{(s,t)}} v_1a_1a_2\cdots a_r$.

Then we have the following result.

Lemma 8.3.7 *In the above procedure, we have* $v_1a_1a_2\cdots a_r \in \mathrm{D}_3(\boldsymbol{A})$ *and* $|Sv_1{}^A| > |Sv_1a_1{}^A|$.

Proof Let $s \in S$. Since $v = v_1v_2v_3 \in \mathrm{D}_3(\boldsymbol{A})$, we have $sv_1{}^A \neq \emptyset$, actually $|sv_1{}^A| = 1$. Notice that $\exists s_r \in S, \forall t \in S, t(a_1a_2\cdots a_r)^A = \{s_r\}$. Therefore, $s(v_1a_1a_2\cdots a_r)^A = (sv_1{}^A)(a_1a_2\cdots a_r)^A = \{s_r\}$ and hence $v_1a_1a_2\cdots a_r \in \mathrm{D}_3(\boldsymbol{A})$. Since $\boldsymbol{A}$ is of *pf*-type and $\{s, t\} \subseteq Sv_1{}^A$, $|Sv_1{}^A| \geq |Sv_1a_1{}^A| + 1$. This completes the proof of the lemma.

Lemma 8.3.8 *Let* $\boldsymbol{A} = (S, X, \delta)$ *be a* D_3*-directable automaton such that* $|S| = n$ *and* $d_3(\boldsymbol{A}) = d_3(n)$. *Then there exists a nondeterministic automaton* $\boldsymbol{B} = (S, Y, \gamma)$ *of pf-type such that* $d_3(\boldsymbol{B}) = d_3(n)$.

Proof Let $u = a_1a_2\cdots a_r \in \mathrm{D}_3(\boldsymbol{A})$ with $|u| = d_3(\boldsymbol{A})$. Since $u \in \mathrm{D}_3(\boldsymbol{A})$, there are $s_r \in S$ and a sequence of partial functions of S into S, $\rho_1, \rho_2, \ldots, \rho_r$ such that $s(a_1a_2\cdots a_i)^A \supseteq \rho_i(\rho_{i-1}(\cdots(\rho_1(s))\cdots))$ for any $s \in S$ and any $i = 1, 2, \ldots, r$. Furthermore, $\rho_r(\rho_{r-1}(\cdots(\rho_1(s))\cdots)) = \{s_r\}$ for any $s \in S$. Now we define the *pf*-type automaton $\boldsymbol{B} = (S, Y, \gamma)$ as follows: (1) $Y = \{b_i \mid i = 1, 2, \ldots, r\}$. Remark that

$b_1, b_2, \ldots, b_r$ are distinct symbols. (2) $sb_i{}^B = \rho_i(s)$ for any $s \in S$ and any $i = 1, 2, \ldots, r$.

Then $\boldsymbol{B}$ is a nondeterministic automaton of *pf*-type. Moreover, it is obvious that $b_1 b_2 \cdots b_r \in \mathrm{D}_3(\boldsymbol{B})$. Suppose that $b_{i_1} b_{i_2} \cdots b_{i_k} \in \mathrm{D}_3(\boldsymbol{B})$ where $i_1, i_2, \ldots, i_k \in \{1, 2, \ldots, r\}$. Then we have $a_{i_1} a_{i_2} \cdots a_{i_k} \in \mathrm{D}_3(\boldsymbol{A})$. Therefore, $k \geq r$ and $r = d_3(\boldsymbol{B})$. This completes the proof of the lemma.

We are now ready to determine an upper bound for $d_3(n)$.

Proposition 8.3.10 *For any* $n \geq 3$, $d_3(n) \leq \sum_{k=2}^{n-1} \binom{n}{k} - \sum_{k=0}^{n-2} \binom{n-2}{k} + n - 1$.

Proof Let $\boldsymbol{A} = (S, X, \delta)$ be a nondeterministic automaton of *pf*-type such that $|S| = n$ and $d_3(n) = d_3(\boldsymbol{A})$. Moreover, let $u = a_1 a_2 \cdots a_r \in \mathrm{D}_3(\boldsymbol{A})$ with $r = d_3(n)$. Let $S_i = S(a_1 a_2 \cdots a_i)^A$ for $i = 1, 2, \ldots, r$. Since $\boldsymbol{A}$ is of *pf*-type and $r = d_3(n) = d_3(\boldsymbol{A})$, $|S| > |S_1| \geq |S_2| \geq \cdots \geq |S_{r-1}| > |S_r| = 1$. Let $S_r = \{s_r\}$. By Lemma 8.3.6, $S, S_1, S_2, \ldots, S_{r-1}$ and S_r are distinct. Moreover, since $|S| > |S_1|$, there exist $s_0, s_1 \in S$ such that $s_0 \neq s_1$ and $s_0 a_1{}^A = s_1 a_1{}^A$. Therefore, we can apply for procedure $\mathcal{Q}_{(s_0,s_1)}$ to $a_1 a_2 \cdots a_r$ if necessary and we can get $a_1 a_2 \cdots a_r \Rightarrow^{\mathcal{Q}_{(s,t)}} v_1 a_1 a_2 \cdots a_r$. Now we apply for procedure $\mathcal{P}$ to $v_1 a_1 a_2 \cdots a_r$ as many times as possible until we cannot apply for procedure $\mathcal{P}$ anymore. Hence we can obtain $v_2 \in \mathrm{D}_3(\boldsymbol{A})$ with $|v_2| \leq 2^{|S|} - |S|$. Then we apply for procedure $\mathcal{Q}_{(s_0,s_1)}$ to v_2. We will continue the same process until we cannot apply for neither procedure $\mathcal{P}$ nor $\mathcal{Q}_{(s_0,s_1)}$ anymore. Notice that this process will be terminated after a finite number of applications of procedures $\mathcal{P}$ and $\mathcal{Q}_{(s_0,s_1)}$. Let $w = c_1 c_2 \cdots c_s, c_i \in X, i = 1, 2, \ldots, s$ be the lastD$_3$-directing word of $\boldsymbol{A}$ which was obtained by the above process. Let $T_i = S(c_1 c_2 \cdots c_i)^A$ for any $i = 1, 2, \ldots, s$. Then $T_i \neq T_j$ for any $i, j = 1, 2, \ldots, s$ with $i < j$ and $\{T_1, T_2, \ldots, T_s\}$ contains at most $n - 2$ elements $T_i, i = 1, 2, \ldots, s$ with $T_i \supseteq \{s_0, s_1\}$. Since $|\{T \subseteq S \mid \{s_0, s_1\} \subseteq T\}| = \sum_{k=0}^{n-3} \binom{n-2}{k}$ and by the above observation, we have $d_3(n) \leq \sum_{k=2}^{n-1} \binom{n}{k} - \sum_{k=0}^{n-2} \binom{n-2}{k} + n - 1$.

Regarding the lower bound for $d_3(n)$, we have the following result.

Proposition 8.3.11 *Let $n \geq 3$. Then $d_3(n) \geq 2^m + 1$ if $n = 2m$ ($d_3(n) \geq 3 \cdot 2^{m-1} + 1$ if $n = 2m + 1$).*

Proof Let $n \geq 3$ and let $S = \{1, 2, \ldots, n\}$. Moreover, let $S_1 = \{1, 2\}$, let $S_2 = \{3, 4\}, \ldots$, let $S_{m-1} = \{2m - 3, 2m - 2\}$ and let $S_m = \{2m - 1, 2m\}$ if $n = 2m$ ($S_m = \{2m - 1, 2m, 2m + 1\}$ if $n = 2m + 1$).

We define the following D$_3$-directable automaton $\boldsymbol{A} = (S, X, \delta)$: (1) $\{T_1, T_2, \ldots, T_k\} = \{\{n_1, n_2, \ldots, n_m\} \mid (n_1, n_2, \ldots, n_m) \in S_1 \times S_2 \times \cdots \times S_m\}$ where $k = 2^m$ if $n = 2m$ ($k = 3 \cdot 2^{m-1}$ if $n = 2m+1$). (2) $T_1 = \{1, 3, 5, \ldots, 2m - 1\}$. (3) $X = \{a, b_1, b_2, \ldots, b_{k-2}, b_{k-1}, c\}$. (4) $1a^A = 2a^A = \{1\}, 3a^A = 4a^A = \{3\}, \ldots, (2m - 3)a^A = (2m - 2)a^A = \{2m - 3\}$ and $(2m - 1)a^A = (2m)a^A = \{2m - 1\}$ if $n = 2m$ ($(2m - 1)a^A = (2m)a^A = (2m + 1)a^A = \{2m - 1\}$ if $n = 2m + 1$). (5) Let $i = 1, 2, \ldots, k - 1$. By ρ_i, we denote a bijection of T_i onto T_{i+1}. Then $tb_i{}^A = \rho_i(t)$ for any $t \in T_i$ and $tb_i{}^A = \emptyset$, otherwise. (6) $tc^A = \{1\}$ for any $t \in T_k$ and $tc^A = \emptyset$, otherwise.

Then it can easily be seen that $ab_1b_2 \cdots b_{k-1}c$ is a D$_3$-directing word of $\boldsymbol{A}$. Notice that $cu, acu, ab_1cu, ab_1b_2cu, \ldots, ab_1b_2 \cdots b_icu \notin$ D$_3(\boldsymbol{A})$ for any $i = 1, 2, \ldots, k - 2$ and any $u \in X^*$, and $T_ia^A = T_1$ for any $i = 1, 2, \ldots, k$. Hence $ab_1b_2 \cdots b_{k-1}c$ is a shortest D$_3$-directing word of $\boldsymbol{A}$. Therefore, $d_3(n) \geq 2^m + 1$ if $n = 2m$ ($d_3(n) \geq 3 \cdot 2^{m-1} + 1$ if $n = 2m + 1$).

Finally, we consider the values $cd_2(n)$ and $d_2(n)$.

Proposition 8.3.12 *For $n \geq 2$, $2^n - n - 1 \leq cd_2(n) \leq d_2(n) < 1 + (2^n - 2)\binom{2^n}{2}$. Remark that $cd_2(1) = d_2(1) = 0$.*

Proof Let $n \geq 2$. It is obvious that $cd_2(n) \leq d_2(n)$. The automaton $\boldsymbol{A}$ in the proof of Theorem 8.3.1 shows that $2^n - n - 1 \leq cd_2(n)$. Let $\boldsymbol{A} = (S, X, \delta)$ be a D$_2$-directable automaton with n states and let $w \in \mathrm{D}_2(\boldsymbol{A})$ such that $|w| = d_2(\boldsymbol{A})$. We construct the following finite automaton $\boldsymbol{B} = (\{T \mid T \subseteq S\}, X, \gamma)$: (1) For any $T \subseteq S$ and $a \in X, Ta^{\boldsymbol{B}} = \bigcup_{t \in T} ta^{\boldsymbol{B}}$.

Then $d_2(\boldsymbol{A}) = d(\boldsymbol{B})$. Notice that $|\{T \mid T \subseteq S\}| = 2^n$. By Proposition 8.1.3, $d(\boldsymbol{B}) < 1 + (2^n - 2)\binom{2^n}{2}$. Thus $2^n - n - 1 \leq cd_2(n) \leq d_2(n) < 1 + (2^n - 2)\binom{2^n}{2}$.

Now we introduce some results on the value of $cd_3(n)$. The results are due to [30].

Let $\boldsymbol{A} = (S, X, \delta)$ be a nondeterministic automaton. A word $w \in X^*$ is said to be a D_3-*merging word* for a pair of two states $s, t \in S$ if $sw^A \cap tw^A \neq \emptyset$.

Proposition 8.3.13 *A complete nondeterministic automaton* $\boldsymbol{A} = (S, X, \delta)$ *is* D_3*-directable if and only if there exists a* D_3*-merging word for any pair of* $s, t \in S$.

Proof ($\Rightarrow$) Obvious from the definition of D_3-directability.

($\Leftarrow$) Let $T \subseteq S$ be a maximal subset of S satisfying the following condition: There exists a word $w \in X^*$ such that $\bigcap_{t \in T} tw^A \neq \emptyset$.

If $T = S$, then $\boldsymbol{A}$ is D_3-directable. Assume $T \neq S$. Let $s \in S \setminus T$ and let $t' \in \bigcap_{t \in T} tw^A$. Consider $sw^A \subseteq S$. Notice that $sw^A \neq \emptyset$ because $\boldsymbol{A}$ is complete. Since there exists a D_3-merging word for the pair $t' \in S$ and $s' \in sw^A$, there exists $w' \in X^*$ such that $t'w'^A \cap s'w'^A \neq \emptyset$. Thus $\bigcap_{t \in T \cup \{s\}} t(ww')^A \neq \emptyset$, which contradicts the maximality of T. Hence $S = T$ and $\boldsymbol{A}$ is D_3-directable.

Lemma 8.3.9 *Let* $\boldsymbol{A} = (S, X, \delta)$ *be a nondeterministic automaton with* $|S| = n$. *If there exists a* D_3*-merging word for* $s, t \in S$, *then there exists a* D_3*-merging word* $w \in X^*$ *for* $s, t \in S$ *such that* $|w| \leq \binom{n}{2}$.

Proof Let $s, t \in S$ and let $w = a_1 a_2 \cdots a_r \in X^*$ where $r \geq 1$ and $a_1, a_2, \cdots, a_r \in X$, be a shortest D_3-merging word for $s, t \in S$.

Then $sw^A \cap tw^A \neq \emptyset$. Let $(s_1, t_1) \in sa_1^A \times ta_1^A, (s_2, t_2) \in s(a_1a_2)^A \times t(a_1a_2)^A, \ldots, (s_{r-1}, t_{r-1}) \in s(a_1a_2\cdots a_{r-1})^A \times t(a_1a_2\cdots a_{r-1})^A$. Suppose $r > \binom{n}{2}$. Then there exist $i, j = 1, 2, \ldots, r-1, i \leq j$ such that $\{s_i, t_i\} = \{s_j, t_j\}$. Assume that $s_i = s_j$ and $t_i = t_j$. Then $s(a_1a_2\cdots a_ia_{j+1}a_{j+2}\cdots a_{r-1}a_r)^A \cap t(a_1a_2\cdots a_ia_{j+1}a_{j+2}\cdots a_{r-1}a_r)^A \neq \emptyset$, i.e. $a_1a_2\cdots a_ia_{j+1}a_{j+2}\cdots a_{r-1}a_r$ is a D$_3$-merging word for $s, t \in S$, which contradicts the minimality of $|w|$. Let $s_i = t_j$ and $t_i = s_j$. We have also $s(a_1a_2\cdots a_ia_{j+1}a_{j+2}\cdots a_{r-1}a_r)^A \cap t(a_1a_2\cdots a_ia_{j+1}a_{j+2}\cdots a_{r-1}a_r)^A \neq \emptyset$, i.e. $a_1a_2\cdots a_ia_{j+1}a_{j+2}\cdots a_{r-1}a_r$ is a D$_3$-merging word for $s, t \in S$, which contradicts the minimality of $|w|$. Consequently, $|w| \leq \binom{n}{2}$.

Lemma 8.3.10 *Let* $\boldsymbol{A} = (S, X, \delta)$ *be a complete* D$_3$*-directable automaton with* $|S| = n$ *and let* T *be a subset of* S *with* $|T| = m$. *Then there exists* $w \in X^*$ *such that* $\bigcap_{t \in T} tw^A \neq \emptyset$ *and* $|w| \leq (m-1)\binom{n}{2}$.

Proof We prove the lemma by induction on $|T|$. For $|T| = 2$, by Lemma 8.3.9, the lemma holds true. Assume that the lemma holds true for $|T| = m-1$. Now let $T = \{t_1, t_2, \ldots, t_{m-1}, t_m\}$. By Lemma 8.3.9, there exists $w_1 \in X^*$ such that $|w_1| \leq \binom{n}{2}$ and $t_{m-1}w_1{}^A \cap t_mw_1{}^A \neq \emptyset$. Let $t'_{m-1} \in t_{m-1}w_1{}^A \cap t_mw_1{}^A$, let $t'_1 \in t_1w_1{}^A$, let $t'_2 \in t_2w_1{}^A, \ldots,$ let $t'_{m-3} \in t_{m-3}w_1{}^A$ and let $t'_{m-2} \in t_{m-2}w_1{}^A$. Then, let $T' = \{t'_1, t'_2, \ldots, t'_{m-2}, t'_{m-1}\}$. By the induction hypothesis, there exists $w_2 \in X^*$ such that $\bigcap_{t' \in T'} t(w_2)^A \neq \emptyset$ and $|w_2| \leq (m-2)\binom{n}{2}$. Then we have $\bigcap_{t \in T} t(w_1w_2)^A \supseteq \bigcap_{t' \in T'} t(w_1w_2)^A \neq \emptyset$. Here $|w_1w_2| = |w_1| + |w_2| \leq \binom{n}{2} + (m-2)\binom{n}{2} = (m-1)\binom{n}{2}$. This completes the proof of the lemma.

Proposition 8.3.14 *Let* $n \geq 1$. *Then* $(n-1)^2 \leq cd_3(n) \leq 1 + (n-2)\binom{n}{2}$.

Proof That $(n-1)^2 \leq cd_3(n)$ follows from Proposition 8.1.2. Let $\boldsymbol{A} = (S, X, \delta)$ be a complete nondeterministic automaton such that D$_3(\boldsymbol{A}) \neq \emptyset$. Let w be a shortest D$_3$-directing word of $\boldsymbol{A}$. If $\boldsymbol{A}$ is a deterministic automaton, then $cd_3(n) \leq d(n)$. Now assume that $\boldsymbol{A}$ is not deterministic, i.e. there exist $a \in X$ and $s \in S$ such that

$|sa^{A}| \geq 2$. Since A is complete, a is a D_3-merging word for some two distinct states in S. Then there exists a subset T of S such that $0 < |T| \leq n-1$ and $sa^A \cap T \neq \emptyset$ for any $s \in S$. By Lemma 8.3.10, there exists $w \in X^*$ such that $\bigcap_{t \in T} tw^A \neq \emptyset$ and $|w| \leq (n-2)\binom{n}{2}$. Therefore, $\bigcap_{s \in S} s(aw)^A \neq \emptyset$ and $|aw| \leq 1 + (n-2)\binom{n}{2}$. Thus $cd_3(n) \leq 1 + (n-2)\binom{n}{2}$.

8.4 Commutative case

In this section, we will deal with commutative nondeterministic automata and related languages alongside the same line as that of the previous section.

A nondeterministic automaton $A = (S, X, \delta)$ is called *commutative* if $s(ab)^A = s(ba)^A$ holds for any $s \in S$ and any $a, b \in X$.

By $\mathcal{L}'^X_{\mathrm{D}}, \mathcal{L}'^X_{\mathrm{CND}(i)}$ and $\mathcal{L}'^X_{\mathrm{ND}(j)}, i, j = 1, 2, 3$, we denote the classes of regular languages of directing words of deterministic commutative automata, of D_i-directing words of complete commutative nondeterministic automata, and of D_j-directing words of commutative nondeterministic automata, respectively.

Since it is easy to verify that $\mathcal{L}'^{\{a\}}_{\mathrm{D}} = \mathcal{L}'^{\{a\}}_{\mathrm{CND}(i)} = \mathcal{L}'^{\{a\}}_{\mathrm{ND}(i)} = \{a^n a^* \mid n \geq 1\}$ for any $i = 1, 2, 3$, we assume that $|X| \geq 2$.

Then we have the following result.

Proposition 8.4.1 *Let $L \subseteq X^*$. Then $L \in \mathcal{L}'^X_{\mathrm{D}}$ if and only if $L \neq \emptyset$, L is regular and commutative, and $L = L \diamond X^*$.*

Proof ($\Rightarrow$) First notice that L is a commutative regular language. By Proposition 8.3.1, $X^*LX^* = L$. Since L is commutative, $X^*LX^* = L \diamond X^*$ and hence $L = L \diamond X^*$.

($\Leftarrow$) The proof can be given in the same way as in Proposition 8.3.1. A nessary task is only to check the following: $\forall u, v \in X^*, [uv] = [vu]$, i.e. $uv \equiv vu(P_L)$.

However, this is obvious from the assumption that L is commutative.

Corollary 8.4.1 $\mathcal{L}'^X_{\mathrm{D}} = \mathcal{L}'^X_{\mathrm{CND(2)}} = \mathcal{L}'^X_{\mathrm{ND(2)}}$.

Proof It is obvious that $\mathcal{L}'^X_{\mathrm{D}} \subseteq \mathcal{L}'^X_{\mathrm{CND(2)}} \subseteq \mathcal{L}'^X_{\mathrm{ND(2)}}$. Therefore, to prove the corollary, it is enough to show that $\mathcal{L}'^X_{\mathrm{D}} = \mathcal{L}'^X_{\mathrm{ND(2)}}$. Assume that $L \in \mathcal{L}'^X_{\mathrm{ND(2)}}$. Then by Lemma 8.3.1, $L = LX^*$ where L is a commutative regular language. Since L and X^* are commutative, $L = LX^* = L \diamond X^*$. By the above proposition, $L \in \mathcal{L}'^X_{\mathrm{D}}$ and hence $\mathcal{L}'^X_{\mathrm{D}} = \mathcal{L}'^X_{\mathrm{ND(2)}}$.

Corollary 8.4.2 *The class $\mathcal{L}'^X_{\mathrm{D}}$ is closed under union and intersection.*

Proof Let $L, M \in \mathcal{L}'^X_{\mathrm{D}}$. Then $L \cup M$ is regular. Moreover, $L \cup M = (L \diamond X^*) \cup (M \diamond X^*) = (L \cup M) \diamond X^*$. Therefore, by the above proposition, $L \cup M \in \mathcal{L}'^X_{\mathrm{D}}$, i.e. $\mathcal{L}'^X_{\mathrm{D}}$ is closed under union.

Let $u \in L$ and $v \in M$. Then $uv \in u \diamond v$ and $uv \in L \cap M$, i.e. $L \cap M \neq \emptyset$. Furthermore, $L \cap M = (L \cap M) \diamond X^*$. Hence $L \cap M \in \mathcal{L}'^X_{\mathrm{D}}$. This means that $\mathcal{L}'^X_{\mathrm{D}}$ is closed under intersection.

Proposition 8.4.2 $\mathcal{L}'^X_{\mathrm{D}} = \mathcal{L}'^X_{\mathrm{CND(1)}} = \mathcal{L}'^X_{\mathrm{CDN(3)}}$.

Proof Let $i = 1, 2$. Then it is obvious that $\mathcal{L}'^X_{\mathrm{D}} \subseteq \mathcal{L}'^X_{\mathrm{CND}(i)}$. Let $\emptyset \neq L \in \mathcal{L}'^X_{\mathrm{CDN}(i)}$. Then there exists a commutative complete automaton $\boldsymbol{A} = (S, X, \delta)$ such that $L = \mathrm{D}_i(\boldsymbol{A})$. Let $xuvy \in \mathrm{D}_i(\boldsymbol{A})$ where $x, u, v, y \in X^*$. Since $s(xuvy)^{\boldsymbol{A}} = s(xvuy)^{\boldsymbol{A}}$ for any $s \in S$, $xvuy \in \mathrm{D}_i(\boldsymbol{A}) = L$. Hence L is commutative. Since $\boldsymbol{A}$ is complete, $X^*\mathrm{D}_i(\boldsymbol{A}) = \mathrm{D}_i(\boldsymbol{A})$, i.e. $X^*L = L$. As L is commutative, $X^*L = L \diamond X^*$. Therefore, $L = L \diamond X^*$. By Proposition 8.4.1, $L \in \mathcal{L}'^X_{\mathrm{D}}$, i.e. $\mathcal{L}'^X_{\mathrm{D}} = \mathcal{L}'^X_{\mathrm{CND}(i)}$.

Now we provide a characterization of the classes $\mathcal{L}'^X_{\mathrm{ND(1)}}$ and $\mathcal{L}'^X_{\mathrm{ND(3)}}$.

Proposition 8.4.3 *Let $L \subseteq X^*$ and let $i = 1, 3$. Then $L \in \mathcal{L}'^X_{\mathrm{ND}(i)}$ if and only if there exist $\emptyset \neq Y \subseteq X$ and a finite nonempty commutative language $L_0 \subseteq Y^*$ such that $L = L_0 \diamond Y^*$.*

Proof (⇒) Let $L \in \mathcal{L}'^{X}_{\mathrm{ND}(i)}$. Then there exists a commutative nondeterministic automaton $\boldsymbol{A} = (S, X, \delta)$ such that $\mathrm{D}_i(\boldsymbol{A}) = L$. Let $Y = \{a \in X \mid \forall s \in S, sa^A \neq \emptyset\}$. It follows from $L \neq \emptyset$ that $Y \neq \emptyset$. Suppose that there exists $a \in X \setminus Y$ such that $a \in alph(L)$. Then there exists $u \in X^*$ with $au \in L$. Then there exists $s \in S$ such that $s(au)^A = \emptyset$. This contradicts the fact that $au \in \mathrm{D}_i(\boldsymbol{A}) = L$. Therefore, $L \subseteq Y^*$. Let L_0 be the shuffle base of L. Recall that the shuffle base is a hypercode and hence it is finite. Moreover, since L is commutative, L_0 is commutative. It is also obvious that $L = L_0^{\diamond}$. On the other hand, since $sv^A \neq \emptyset$ for any $s \in S$ and any $v \in Y^*, Y^*L_0 \subseteq L$ and in fact $Y^*L_0 = L$. As L_0 is commutative, $L = Y^*L_0 = L_0 \diamond Y^*$.

(⇐) By Proposition 8.4.1, there exists a commutative nondeterministic automaton $\boldsymbol{A} = (S, Y, \delta)$ such that $L = \mathrm{D}_i(\boldsymbol{A})$. We will construct the commutative nondeterministic automaton $B = (S, X, \gamma)$ as follows:

(1) $sa^B = sa^A$ for any $s \in S$ and any $a \in Y$. (2) $sa^B = \emptyset$ for any $s \in S$ and any $a \in X \setminus Y$.

Then obviously $L = \mathrm{D}_i(\boldsymbol{B})$ and $\boldsymbol{B}$ is commutative because $s(ab)^B = s(ab)^A = s(ba)^A = s(ba)^B$ for any $s \in S$ and any $a, b \in Y$ and, $s(ab)^B = \emptyset = s(ba)^B$ for any $s \in S$ and any $a \in X \setminus Y, b \in X$. Therefore, $L \in \mathcal{L}'^{X}_{\mathrm{ND}(1)}$.

Corollary 8.4.3 $\mathcal{L}'^{X}_{\mathrm{ND}(1)} = \mathcal{L}'^{X}_{\mathrm{ND}(3)}$ *and* $\mathcal{L}'^{X}_{\mathrm{D}} \subset \mathcal{L}'^{X}_{\mathrm{ND}(1)}$.

Proof That $\mathcal{L}'^{X}_{\mathrm{ND}(1)} = \mathcal{L}'^{X}_{\mathrm{ND}(3)}$ is obvious from Proposition 8.4.3. First, it is obvious that $\mathcal{L}'^{X}_{\mathrm{D}} \subseteq \mathcal{L}'^{X}_{\mathrm{ND}(1)}$. Let $a \in X$ and let $|X| > 1$. Then, by Proposition 8.4.3, $a^+ \in \mathcal{L}'^{X}_{\mathrm{ND}(1)}$. However, $a^+ \notin \mathcal{L}'^{X}_{\mathrm{D}}$. Hence $\mathcal{L}'^{X}_{\mathrm{D}} \subset \mathcal{L}'^{X}_{\mathrm{ND}(1)}$.

Corollary 8.4.4 $\mathcal{L}'^{X}_{\mathrm{ND}(1)}$ *is neither closed under union nor closed under intersection.*

Proof Let $|X| > 1$. Then $a^+, b^+ \in \mathcal{L}'^{X}_{\mathrm{ND}(1)}$ where $a, b \in X, a \neq b$. By Proposition 8.4.3, $a^+ \cup b^+ \in \mathcal{L}'^{X}_{\mathrm{ND}(1)}$, i.e. $\mathcal{L}'^{X}_{\mathrm{ND}(1)}$ is not closed under intersection. Now consider $a^+ \cap b^+ = \emptyset$ and hence $a^+ \cap b^+ \notin \mathcal{L}'^{X}_{\mathrm{ND}(1)}$. Consequently, $\mathcal{L}'^{X}_{\mathrm{ND}(1)}$ is not closed under intersection.

Here are the inclusion relations for the 7 classes of commutative regular languages.

$$\mathcal{L}'^{X}_{\mathrm{ND}(1)} = \mathcal{L}'^{X}_{\mathrm{ND}(3)}$$

$$\mathcal{L}'^{X}_{\mathrm{D}} = \mathcal{L}'^{X}_{\mathrm{ND}(2)} = \mathcal{L}'^{X}_{\mathrm{CND}(1)} = \mathcal{L}'^{X}_{\mathrm{CND}(2)} = \mathcal{L}'^{X}_{\mathrm{CND}(3)}$$

Figure 8.2: Commutative case

We will consider the shortest directing words of commutative nondeterministic automata.

Let $i = 1, 2, 3$ and let $n \geq 1$. Then $cd_{com(i)}(n)$ denotes the value $max\{d_i(\boldsymbol{A}) \mid \boldsymbol{A} = (S, X, \delta)$: commutative, $\boldsymbol{A} \in \mathbf{CDir}(i)$ and $|S| = n\}$.

Notice that in the definitions of $d_{com(i)}(n)$ and $cd_{com(i)}(n)$, X ranges over all finite nonempty alphabets.

Example 8.4.1 Let $n \geq 2$ and let $\boldsymbol{A} = (\{1, 2, \ldots, n\}, \{a\}, \delta)$ be the nondeterministic automaton such that $1^A = \{1, 2\}, ia^A = \{i + 1\}$ for any $i = 2, 3, \ldots, n-1$ and $na^A = \{1\}$. Notice that $\boldsymbol{A}$ is commutative. Then it can be verified that $i(a^{2n-2})^A = \{1, 2, \ldots, n\}$ for any $i = 1, 2, \ldots, n$ but $2(a^k)^A = \{2 + k\}$ for any $k = 1, 2, \ldots, n - 2$ and $2(a^k)^A = \{1, 2, \ldots, k - n + 2\}$ for any $k = n - 1, n, \ldots, 2n - 3$. This means that a^{2n-2} is the shortest D_2-directing word of $\boldsymbol{A}$. i.e. $d_2(\boldsymbol{A}) = 2n - 2$. Consequently, $d_{com(2)}(n) \geq 2n - 2$.

Lemma 8.4.1 *Let* $\boldsymbol{A} = (S, X, \delta)$ *be a commutative nondeterministic automaton. Let* $i = 1, 3$*, let* $L_i = \mathrm{D}_i(\boldsymbol{A})$ *and let* $Y = alph(L_i)$*. If* $\mathrm{D}_i(\boldsymbol{A}) \neq \emptyset$*, then* $sa^A \neq \emptyset$ *for any* $s \in S$ *and any* $a \in Y$*.*

Proof Let $s \in S$ and let $a \in Y$. Since $a \in alph(L_i)$, there exist $u, v \in X^*$ such that $uav \in L_i$ and hence $auv \in L_i$. Therefore, $s(auv)^A = S(auv)^A \neq \emptyset$ for any $s \in S$. Thus $sa^A \neq \emptyset$

Lemma 8.4.2 *Let $n \geq 1$ and let $i = 1, 2, 3$. Then we have $d_{com(i)}(n)$ $= cd_{com(i)}(n)$ for any $n \geq 1$.*

Proof It is obvious that $cd_{com(i)}(n) \leq d_{com(i)}(n)$. First consider the cases $i = 1, 3$. Let $\boldsymbol{A} = (S, X, \delta)$ be a commutative nondeterministic automaton such that $d_i(\boldsymbol{A}) = d_{com(i)}(n)$. If $\boldsymbol{A}$ is complete, then $d_{com(i)}(n) = cd_{com(i)}(n)$. Now assume that $\boldsymbol{A}$ is not complete. Let $L_i = \mathrm{D}_i(\boldsymbol{A})$ and let $Y = alph(L_i)$. By Lemma 8.4.1, $L_i \subseteq Y^*$. We construct the following nondeterministic automaton $\boldsymbol{B} = (S, Y, \gamma)$: $sa^B = sa^A$ for any $s \in S$ and any $a \in Y$.

Notice that $\boldsymbol{B}$ is complete, commutative, $|S| = n$ and $d_i(\boldsymbol{B}) = d_{com(i)}(n)$. Hence $d_{com(i)}(n) = cd_{com(i)}(n)$.

Now we consider the case $i = 2$. For $n = 1, 2$, it is obvious that $d_{com(2)}(n) = cd_{com(2)}(n)$. Hence we assume that $n > 2$. Let $\boldsymbol{A} = (S, X, \delta)$ be a commutative nondeterministic automaton such that $d_2(\boldsymbol{A}) = d_{com(2)}(n)$. Let $u = a_1 a_2 \cdots a_r$ be a shortest D_2-directing word of $\boldsymbol{A}$. If $Su^A \neq \emptyset$, then we construct the following commutative nondeterministic automaton $\boldsymbol{B} = (S, Y, \gamma)$ where $Y = \{a \in X \mid \forall s \in S, sa^A \neq \emptyset\}$ and $sa^B = sa^A$ for any $s \in S$ and any $a \in Y$. By Lemma 8.4.1, $\boldsymbol{B}$ is complete. Furthermore, u is a shortest word in $\mathrm{D}_2(\boldsymbol{B})$, i.e. $d_{com(2)}(n) \leq d_2(\boldsymbol{B})$. Hence $d_{com(2)}(n) \leq cd_{com(2)}(n)$. Therefore, $d_{com(2)}(n) = cd_{com(2)}(n)$.

Now assume $Su^A = \emptyset$. Notice that $Su^A = \bigcup_{s \in S} su^A \supseteq S(vu)^A = S(uv)^A$ for any $u, v \in X^*$. Hence $Sa_1^A \supseteq S(a_1a_2)^A \supseteq \cdots \supseteq S(a_1a_2 \cdots a_{r-1})^A \supseteq S(a_1a_2 \cdots a_r)^A = \emptyset$. If $S(a_1a_2 \cdots a_j)^A = S(a_1a_2 \cdots a_{j+1})^A$ for some $j = 1, 2, \ldots, n-1$, then $S(a_1a_2 \cdots a_{j-1}a_{j+1} \cdots a_r)^A = \emptyset$ and $a_1a_2 \cdots a_{j-1}a_{j+1} \cdots a_r \in \mathrm{D}_2(\boldsymbol{A})$, which contradicts the assumption that $a_1a_2 \cdots a_r$ is a shortest D_2-directing word of $\boldsymbol{A}$. Therefore, $Sa_1^A \supset S(a_1a_2)^A \supset \cdots \supset S(a_1a_2 \cdots a_{r-1})^A \supset S(a_1a_2 \cdots a_r)^A = \emptyset$. Since $|S| = n, r \leq n$ and hence $d_{com(2)}(n) \leq n$. Recall that $2n - 2 \leq d_{com(2)}(n)$ (see Example 8.4.1). This yealds a contradiction. Thus $d_{com(2)}(n) = cd_{com(2)}(n)$.

Theorem 8.4.1 *For any $n \geq 1, d_{com(1)}(n) = cd_{com(1)}(n) = n - 1$.*

Proof Let $\boldsymbol{A} = (S, X, \delta)$ be a commutative nondeterministic automaton with $|S| = n$. Assume that $\mathrm{D}_1(\boldsymbol{A}) \neq \emptyset$. Let $u = a_1 a_2 \cdots a_r$

be a shortest word in $D_1(\boldsymbol{A})$ where $r \geq 1$ and $a_1, a_2, \cdots, a_r \in X$. Then $Sa_1^A \supseteq S(a_1a_2)^A \supseteq \cdots \supseteq S(a_1a_2\cdots a_r)^A$ where $|S(a_1a_2\cdots a_r)^A| = 1$. If $S(a_1a_2\cdots a_j)^A = S(a_1a_2\cdots a_{j+1})^A$ for some $j = 1, 2, \ldots, n-1$, then $|S(a_1a_2\cdots a_{j-1}a_{j+1}\cdots a_r)^A| = 1$ and $a_1a_2\cdots a_{j-1}a_{j+1}\cdots a_r \in D_1(\boldsymbol{A})$, which contradicts the minimality of $|u|$. Therefore, $Sa_1^A \supset S(a_1a_2)^A \supset \cdots \supset S(a_1a_2\cdots a_r)^A$. Since $S \supset Sa_1^A, |S| = n$ and $|S(a_1a_2\cdots a_r)^A| = 1$, we have $r \leq n-1$ and $d_{com(1)}(n) = cd_{com(1)}(n) \leq n-1$.

Consider the automaton $\boldsymbol{B} = (\{1, 2, \ldots, n\}, \{a\}, \delta)$ where $ia^B = \{i+1\}$ for $I = 1, 2, \ldots, n-1$ and $na^B = n$. Then it is obvious that $d_i(\boldsymbol{A}) = n-1$. Hence $d_{com(1)}(n) = cd_{com(1)}(n) \geq n-1$.

This completes the proof of the proposition.

Example 8.4.2 Let $n \geq 3$ and let $\boldsymbol{A} = (\{1, 2, \ldots, n\}, \{a\}, \delta)$ be the nondeterministic automaton such that $1^A = \{2, 3\}, ia^A = \{i+1\}$ for any $i = 2, 3, \ldots, n-1$ and $na^A = \{1\}$. Then $\boldsymbol{A}$ is commutative and complete.

Lemma 8.4.3 *Let* $\boldsymbol{A}$ *be the complete nondeterministic automaton in Example 8.4.2. Then* $d_2(\boldsymbol{A}) = (n-1)^2 + 1$.

Proof It can easily be verified that $ia^{(n-1)^2+1} = \{1, 2, 3, \ldots, n\}$ for any $i = 1, 2, 3, \ldots, n$ but $2a^{(n-1)^2} = \{1, 3, 4, \ldots, n-1, n\} \neq \{1, 2, 3, \ldots, n\}$. Moreover, $1a^{(n-1)^2} = \{1, 2, 3, \ldots, n\}$. Therefore, $a^{(n-1)^2+1} \in D_2(\boldsymbol{A})$ but $a^{(n-1)^2} \notin D_2(\boldsymbol{A})$. Since $D_2(\boldsymbol{A})X^* = D_2(\boldsymbol{A})$, $d_2(\boldsymbol{A}) = (n-1)^2 + 1$.

Lemma 8.4.4 *Let* $\boldsymbol{A}$ *be the complete nondeterministic automaton in Example 8.4.2. Then* $d_3(\boldsymbol{A}) = n(n-3) + 3$.

Proof It can be easily seen that $D_2(\boldsymbol{A}) \neq \emptyset$. Let $m = d_3(\boldsymbol{A})$. Then we have $\bigcap_{i=1}^{n} i(a^m)^A \neq \emptyset$. Let $j \in \bigcap_{i=1}^{n} i(a^m)^A$. If $j = 1, 2$, then

$\{n, 1\} \cap (\bigcap_{i=1}^{n} i(a^{m-1})^A) \neq \emptyset$ and $d_3(\boldsymbol{A}) < m$, which is a contradiction.

If $j > 3$, then we have $j - 1 \in \bigcap_{i=1}^{n} i(a^{m-1})^A$ and $d_3(\boldsymbol{A}) < m$, which is a contradiction. Consequently, $j = 3$. It can easily be verified that $3 \in i(a^{n(n-3)+3})^A$ for any $i = 1, 2, \ldots, n$ but $3 \notin 2(a^{n(n-3)+2})^A$. Hence $a^{n(n-3)+3} \in \mathrm{D}_3(\boldsymbol{A})$ but $a^{n(n-3)+2} \notin \mathrm{D}_3(\boldsymbol{A})$. Since $\boldsymbol{A}$ is complete, $\mathrm{D}_3(\boldsymbol{A})X^* = \mathrm{D}_3(\boldsymbol{A})$. Therefore, $d_3(\boldsymbol{A}) = n(n-3) + 3$.

Proposition 8.4.4 *Let $n \geq 2$. Then $(n-1)^2 + 1 \leq cd_{com(2)}(n) = d_{com(2)}(n) \leq 2^n - 2$. For $n = 1$, $cd_{com(2)}(1) = d_{com(2)}(1) = 0$.*

Proof For $n = 1, 2$, obviously the proposition holds. Let $n \geq 3$ and let $\boldsymbol{A} = (S, X, \delta) \in \mathbf{CDir}(2)$ with $|S| = n$. We construct the following complete nondeterministic automaton $\boldsymbol{B} = (\{T \subseteq S \mid T \neq \emptyset\}, X, \gamma)$:

$$aT^B = \bigcap_{t \in T}^{n} ta^A \text{ for any } T, \emptyset \neq T \subseteq S \text{ and any } a \in X.$$

Then it is obvious that $\boldsymbol{B}$ is complete, $|\{T \subseteq S \mid T \neq \emptyset\}| = 2^n - 1$, and $d_2(\boldsymbol{A}) = d_1(\boldsymbol{B})$. By Theorem 8.4.1, $d_1(\boldsymbol{B}) \leq 2^n - 2$ and $d_2(\boldsymbol{A}) \leq 2^n - 2$. Hence $cd_2(n) = d_2(n) \leq 2^n - 2$. Now we show that $(n-1)^2 + 1 \leq cd_2(n) = d_2(n)$. Let $\boldsymbol{A}$ be the complete nondeterministic automaton in Example 8.4.2. By Theorem 8.4.1, we have $d_2(\boldsymbol{A}) = (n-1)^2 + 1$. Thus $(n-1)^2 + 1 \leq cd_2(n)$.

Proposition 8.4.5 *Let $n \geq 2$. Then $n^2 - 3n + 3 \leq cd_{com(3)}(n) = d_{com(3)}(n) \leq 1 + (n-2)\binom{n}{2}$. For $n = 1$, $cd_{com(3)}(1) = d_{com(3)}(1) = 0$.*

Proof For $n = 1, 2$, obviously the proposition holds. Let $n \geq 3$ and let $\boldsymbol{A} = (S, X, \delta) \in \mathbf{CDir}(3)$ with $|S| = n$. By Proposition 8.3.14, $d_{com(3)}(\boldsymbol{A}) \leq 1 + (n-2)\binom{n}{2}$. Hence $cd_{com(3)}(n) = d_3(n) \leq 1 + (n-2)\binom{n}{2}$. On the other hand, consider the complete nondeterministic automaton $\boldsymbol{A}$ in Example 8.4.2. By Lemma 8.4.4, $n^2 - 3n + 3 = cd_3(\boldsymbol{A})$. Hence $n^2 - 3n + 3 \leq cd_{com(3)}(n) = d_{com(3)}(n)$.

Bibliography

[1] T. Araki and N. Tokura, Decision problems for regular expressions with shuffle and shuffle closure operators, Transactions of Institute of Electronics and Communication Engineers of Japan, J64-D (1981), 1069-1073.

[2] T. Araki and N. Tokura, Flow languages equal recursively enumerable languages, Acta Informatica 15 (1981), 209-217.

[3] B.H. Barnes, Groups of automorphisms and set of equivalence classes of input for automata, Journal of the Association for Computing Machinery 12 (1965), 561-565.

[4] B.H. Barnes, On the groups of automorphisms of strongly connected automata, Mathematical Systems Theory 4 (1970), 289-294.

[5] R. Bayer, Automorphism groups and quotients of strongly connected automata and monadic algebras, Report No. 204, Department of Computer Science, University of Illinois (1966).

[6] Z. Bavel, Structure and transition-preserving functions of finite automata, Journal of the Association for Computing Machinery 15 (1968), 135-158.

[7] B. Berard, Literal shuffle, Theoretical Computer Science 51 (1987), 281-299.

[8] J. Berstel and D. Perrin, *Theory of Codes*, Academic Press, New York, 1985.

[9] G. Birkhoff, *Lattice Theory*, Providence, R.I. AMS, 1967.

[10] H.V. Burkhard, Zum Längenproblem homogener experimente an determinierten und nicht-deterministischen automaten, Elektronische Informationsverarbeitung und Kybernetik, EIK 12 (1976), 301-306.

[11] J. Černý, Poznámka k homogénym experimentom s konečinými automatami, Matematicko-fysikalny Časopis SAV 14 (1964), 208-215.

[12] A. Clifford and G.B. Preston, *The Algebraic Theory of Semigroups I*, Mathematical Survey of the American Mathematical Society 7, Providence, R.I., 1961 (1964), 208-215.

[13] A. Clifford and G.B. Preston, *The Algebraic Theory of Semigroups II*, Mathematical Survey of the American Mathematical Society 7, Providence, R.I., 1967 (1964), 208-215.

[14] C. Câmpeanu, K. Salomaa and S. Vágvölgyi, Shuffle quotient and decompositions, Lecture Notes in Computer Science 2295 (Springer) (2002), 186-196.

[15] S. Eilenberg, *Automata, Languages and Machines, Vol A*, Academic Press, New York, 1974.

[16] S. Eilenberg, *Automata, Languages and Machines, Vol. B*, Academic Press, New York, 1976.

[17] A. C. Fleck, Isomorphism groups of automata, Journal of the Association for Computing Machinery 9 (1962), 469-476.

[18] A. C. Fleck, On the automorphism group of an automaton, Journal of the Association for Computing Machinery 12 (1965), 566-569.

[19] S. Ginsburg, *The Mathematical Theory of Context-Free Languages*, McGraw Hill, New York, 1975.

[20] A. Ginzburg, *Algebraic Theory of Automata*, Academic Press, 1968.

[21] J. W. Grzymala-Busse, On the strongly related automata, Foundations of Control Engineering 3 (1978), 3-8.

[22] M. Hall, *The Theory of Groups*, Macmillan, New York, 1959.

[23] J. Hartmanis and R. E. Stearns, *Algebraic Structure Theory of Sequential Machines*, Prentice-Hall, 1966.

[24] J.E. Hopcroft and J.D. Ullman, *Introduction to Automata Theory, Languages and Computation*, Addison-Wesley, Reading MA, 1979.

[25] B. Imreh and M. Ito, On some special classes of regular languages, in *Jewels are Forever* (edited by J. Karhumäki et al.) (1999) (Springer, New York), 25-34.

[26] B. Imreh and M. Ito, On regular languages determined by nondeterministic directable automata, Acta Cybernetica, to appear.

[27] B. Imreh, M. Ito and M. Katsura, On shuffle closures of commutative regular languages, in *Combinatorics, Complexity and Logic* (edited by D.S. Bridges et al.) (1996) (Springer, Singapore), 276-288.

[28] B. Imreh, M. Ito and M. Steinby, On commutative directable nondeterministic automata, in *Grammars and Automata for Strings: From Mathematics and Computer Science to Biology, and Back* (edited by C. Martin-Vide et al.) (2003) (Taylor and Francis, London), 141-150.

[29] B. Imreh and M. Steinby, Some remarks on directable automata, Acta Cybernetica 12 (1995), 23-36.

[30] B. Imreh and M. Steinby, Directable nondeterministic automata, Acta Cybernetica 14 (1999), 105-115.

[31] M. Ito, A representation of strongly connected automata, Systems-Computers-Controls 7 (1976), 29-35.

[32] M. Ito, Strongly connected group-matrix type automata whose orders are prime, Transactions of Institute of Electronics and Communication Engineers of Japan E59 No. 9 (1976), 6-10.

[33] M. Ito, Generalized group-matrix type automata, Transactions of Institute of Electronics and Communication Engineers of Japan E59 No. 11 (1976), 9-13.

[34] M. Ito, A representation of strongly connected automata and its Applications, Journal of Computer and Systems Science 17 (1978), 65-80.

[35] M. Ito, Input sets of strongly connected automata, Fundamentals of Computation Theory, Band 2 (1979), 187-192.

[36] M. Ito, Some classes of automata as partially ordered sets, Mathematical Systems Theory 15 (1982), 357-370.

[37] M. Ito, On small categories whose sets of objects form lattices, Colloquia Mathematica János Bolyai Vol. 39 (North Holland, Amsterdam) (1985), 83-96.

[38] M. Ito, Shuffle decomposition of regular languages, Journal of Universal Computer Science 8 (2002), 257-259.

[39] M. Ito and J. Duske, On cofinal and definite automata, Acta Cybernetica 6 (1983), 181-189.

[40] M. Ito, L. Kari and G. Thierrin, Insertion and deletion closure of languages, Theoretical Computer Science 183 (1997), 3-19.

[41] M. Ito, L. Kari and G. Thierrin, Shuffle and scattered deletion closure of languages, Theoretical Computer Science 245 (2000), 115-134.

[42] M. Ito and M. Katsura, On transitive cofinal automata, Mathematical Aspects of Natural and Formal Languages (1994) (World Scientific, Singapore), 173-199.

[43] M. Ito, C.M. Reis and G. Thierrin. Adherence in finitely generated free monoids. Congressus Numerantium 95 (1993), 37-45.

[44] M. Ito, P.V. Silva, Sremarks on deletion, scattered deletions and related operations on languages, in *Semigroups and Applications* (edited by J. J.M. Howie et al.) (1998) (World Scientific, Singapore), 97-105.

[45] M. Ito and R. Sugiura, n-Insertion on languages, to appear in Aspects of Molecular Computing.

[46] M. Ito and G. Tanaka, Dense property of initial literal shuffles, International Journal of Computer Mathematics 34 (1990), 161-170.

[47] M. Ito, G. Thierrin and S.S. Yu, Shuffle-closed languages, Publicationes Mathematicae of Debrecen 46/3-4 (1995),1-21.

[48] M. Ito and K. Tsuji, Some shortest directing words of nondeterministic directable automata, submitted.

[49] K. Iwama, Universe problem for unrestricted flow languages, Acta Informatica 19 (1983), 85-96.

[50] M. Jantzen, Extending regular operations with iterated shuffle, Theoretical Computer Science 38 (1985), 223-247.

[51] J. Jedrzejowicz, Nesting of shuffle closure is important, Information Processing Letters 25 (1987), 363-367.

[52] J. Jedrzejowicz, Infinite hierarchy of shuffle expressions over a finite alphabet, Information Processing Letters 36 (1990), 13-17.

[53] J. Jedrzejowicz, Undecidability results for shuffle languages, Journal of Automata, Languages and Combinatorics 1 (1996), 147-159.

[54] L. Kari. *On insertion and deletion in formal languages*, PhD Thesis, University of Turku, 1991.

[55] M. Katsura, Automorphism groups and factor automata of strongly connected automata, Journal of Computer and Systems Science 36 (1988), 25-65.

[56] M. Lothaire, *Combinatorics on Words*, Addison-Wesley, Reading, 1983.

[57] Y. Masunaga, N. Noguchi and J. Oizumi, A structure theory of automata characterized by group, Journal of Computer and Systems Science 7, 300-305 (1973).

[58] J.-E. Pin, Sur les mots synchronisants dans un automata fini, Elektronische Informationsverarbeitung und Kybernetik, EIK 14 (1978), 297-303.

[59] J.-E. Pin, Sur un cas particulier de la conjecture de Cerny, Automata, Lecture Notes in Computer Science 62 (Springer) (1979), 345-352.

[60] I. Rystsov, Exact linear bound for the length of reset words in commutative automata, Publicationes Mathematicae of Debrecen 48 (1996), 405-409.

[61] I. Rystsov, Reset words for commutative and solvable automata, Theoretical Computer Science 172 (1997), 273-279

[62] A. Salomaa, *Formal languages*, Academic Press, New York, 1973.

[63] Y. Shibata, On the structure of an automaton and automorphisms, Systems-Computers-Controls 3 (1972), 10-15.

[64] H. J. Shyr and G. Thierrin, Hypercodes, Information and Control 24 (1974), 45-54.

[65] H. J. Shyr, *Free monoids and Languages*, Lecture Notes, National Chung-Hsing University, Hon Min Book Company, Taichung, Taiwan, 1991.

[66] G. Thierrin, Hypercodes, right convex languages and their syntactic monoids, Proceedings of the American Mathematical Society 83 (1981), 255-258.

[67] C.A. Trauth, Jr, Group-type automata, Journal of the Association for Computing Machinery 13, 170-175 (1966).

[68] G.P. Weeg, The structure of an automaton and its operation-preserving transformation group, Journal of the Association for Computing Machinery 9 (1962), 345-349.

[69] G.P. Weeg, The automorphism group of direct product of strongly related automata, Journal of the Association for Computing Machinery 12 (1965), 187-195.

[70] D. Wood, *Theory of Computation*, Harper & Row, Publishers, Inc., New York, 1987.

Index

www.ingramcontent.com/pod-product-compliance
Lightning Source LLC
Chambersburg PA
CBHW050640190326
41458CB00008B/2354
* 9 7 8 9 8 1 0 2 4 7 2 7 0 *